The R.A.M.S. Library of Alchemy

Volume 45

Potpourri of Alchemy
Part 2

Compiled by
Hans W. Nintzel

R.A.M.S. Publishing Company

Potpourri of Alchemy
Part 2

Compiled by
Hans W. Nintzel

Produced by

Restorers of Alchemical Manuscripts Society

R.A.M.S. Publishing Company

R.A.M.S. Publishing Company
117 Rutherford Lane
Stuarts Draft VA 24477

R.A.M.S. Library of Alchemy Volume 45:
Potpourri of Alchemy Part 2
Copyright © 2015 R.A.M.S. Publishing Company

First Edition 2015

ISBN-13 978-1519598622

ISBN-10 1519598629

Image Processing by Philip N. Wheeler

This book is sold for informational purposes only. Neither the publisher nor the editor shall be held accountable for the use or misuse of the information in this book.

Printed in the United States of America

Table of Contents

Dedicated to Hans W. Nintzel,

American Alchemist

and

Founder of the

Restorers of Alchemical Manuscripts Society

(R.A.M.S.)

Disclaimer

Liability: The publisher does not warrant or assume any legal liability or responsibility for the accuracy, completeness, or usefulness of any information, apparatus, product, or process disclosed. The publisher makes no representation as to the accuracy or completeness of the contents of this book and specifically disclaims any implied warranty of merchantability or fitness for a particular purpose. No warranty may be created or extended by written sales materials or sales representatives. You should obtain professional consultation where appropriate. The publisher shall not be liable for any loss of profit or other commercial or personal damages, including but not limited to special, incidental, consequential, or other damages.

PREFACE

In assembling the R.A.M.S. materials, I have come across a good many pieces of information that are not large enough to be a "unit" or "book" in itself. Nonetheless, the data is of enough value to warrant its saving and publication. These data are often pieces of information vouchsafed to me over the last twenty years, or are bits collected here and there. Or are short tracts from various documents deemed of value. These have been collected and are being presented under the heading "POTPOURRI ALCHEMIA".

In these volumes will be found these minutæ & reproductions of two of the scarce "Golden Manuscripts" produced by my teacher, Frater Albertus. This collection, then, represents personal instruction, data no longer obtainable and those treatises that hold some promise for alchemical experimenters. We hope this mélange will be of value to alchemical students and that, perhaps, some insight will be obtained from them.

Hans W. Nintzel

March 1990

Introduction

Philip N. Wheeler

"Potpourri Alchemia, An Alchemical Anthology
containing such tracts as: Joducus, Mutus Liber,
Efferarius, recipes of all kinds, etc. Produced by
R.A.M.S. 1990," is a compilation of short works on
Alchemy. This volume contains valuable information
of use to the student of Alchemy. The section titled
"Recipes" is of particular interest: it contains
notes from the period of time when Hans was studying
under Frater Albertus at the Paracelsus Research
Society laboratory in Salt Lake City, Utah.

"Alchemical Symbols," Vol. 21 in the R.A.M.S.
Library of Alchemy, is highly recommended to the
student of Alchemy.

Oil of Iron

Since it is difficult, in fact almost
impossible to obtain pure metallic Iron, it is
preferable to commence the procedure by using C. P.
(Chemically Pure) Ferrous Sulphate Crystals (Fe SO_4
$7H_2O$), the Vitriol Salts of this metal. (Use Baker's
Analyzed C. P. Ferrous Sulphate obtainable from any
Chemical Reagent Supply House). These Salts have
already been purified by several successive washings
and re-crystallization.

Using a Number 3 Porcelain Coors Crucible
(glazed inside and outside) level full with the
Ferrous Sulphate Crystals (five) calcine the Salts
in the electric muffle furnace. Place the crucible
containing the SALTS in the *cold* furnace and then
raise the temperature of the furnace to 1,750 C.
Continue the calcination until all fuming ceases and
the iron is brought completely to a state of
oxidation. This usually takes about two hours. A
good supply of air should have access to the
furnace. Remove the crucible and allow to cool to
nearly room temperature. Transfer the oxide sponge

13

to a glass mortar and triturate with a glass pestle to a fine powder.

Weigh out 14 grams (approximately one half ounce avoirdupois) of the powdered iron oxide and transfer to a Pyrex Beaker of about 800 ml. capacity and digest on the hot plate with 120 ml. of 6 normal Hydrochloric Acid until all the Iron oxide is in solution, adding more acid if necessary to accomplish solution. Add 20 ml. of C. P. Nitric Acid (Specific Gravity 1.42) and boil for five minutes to bring all the iron to the Ferric State. (Note: It cannot be assumed that the Iron Oxide product of the calcination is in a uniform condition of oxidation). Cool for a few minutes and filter the solution through a fast filter paper with distilled water and allowing the washings to drain through this filter paper into the Beaker containing the main portion of the filtrate. Discard the filter paper and any residue remaining in it.

Make up the volume of the Solution in the Pyrex Beaker to 1,000 ml. Neutralize the Solution with 6 normal Sodium Hydroxide Solution (To make: Add to 250 gm. of NaOH, purified by alcohol, enough distilled water to make the volume 1000 ml.) thus precipitate the Iron as Ferric Hydroxide, then add a slight excess of the 6 normal Sodium Hydroxide Solution. The solution containing the Iron should be stirred vigorously with a glass stirring rod during the preceding operation. Still stirring vigorously, add 6 normal Hydrochloric Acid cautiously, using a dropper until the Solution is neutral or to the barest trace of acid reaction to litmus paper.

Make up the volume of the contents of the Pyrex Beaker to about 1,900 ml. with distilled water. Stir well and then allow the Ferric Hydroxide precipitate to settle for about an hour. Decant the clear Supernatant Solution. Again bring the volume of the contents of the Beaker to about 1,900 ml., stirring well as the distilled water is added, and allow to settle. Repeat this washing by decantation three or four times. (See Note 1.)

Heat the Solution, containing the washed Ferric Hydroxide precipitate, after the last decantation and without further addition of distilled water, to near boiling and filter through a *hardened* filter paper (Whatman No. 52 or No. 54) using vacuum with a Buchner Vacuum Filter Funnel and a Vacuum Filter Flask and finally washing any Ferric Hydroxide precipitate adhering to the wall of the Beaker into the Buchner Filter Funnel by means of a fine stream of distilled water from the wash bottle.

After the filtrate has been sucked through the Ferric Hydroxide cake into the Filter Flask, wash the Ferric Hydroxide cake in the Buchner with three consecutive 100 ml. portions of hot distilled water, taking great care to seal up with the end of a glass stirring rod, all cracks which may develop in the Ferric Hydroxide cake, *as they appear*.

Maintain the vacuum in the Filter Flask after the last washing until the Ferric Hydroxide cake is free of excess water and is firm and solid. Remove the cake and filter paper by inserting the blade of a small spatula between the cake and the porcelain wall of the Buchner and running it around the periphery of the cake. Detach the Buchner Funnel from the Vacuum Filter Flask and invert the Buchner over a porcelain dish of sufficient diameter to receive the cake and filter paper and then jar these from the Buchner into the porcelain dish. Peel off and remove the hardened filter paper (this may be washed and used again) from the cake. Remove carefully any Ferric Hydroxide still adhering to the Buchler Funnel and add it to the main portion of the cake in the porcelain dish. Discard the filtrate and wash waters from the vacuum filtration.

Place the porcelain dish, containing the Ferric Hydroxide cake in the drying oven and dry at 225 F. for about three hours. At intervals, as the drying proceeds, break up the cake in the dish with spatula or knife blade into smaller and smaller pieces until, finally, it has been chopped up into small pieces about the size of a match head. This will ensure proper drying. When the precipitate is finally quite dry, remove from the drying oven and allow to cool. Then transfer the dried material to a glass mortar and triturate it to a fine powder with the glass pestle.

Transfer the fine powder to a Pyrex Beaker of 400 ml. capacity. Add sufficient fuming C. P. Hydrochloric Acid (specific gravity 1.18), to dissolve the powdered iron oxide (approximately 50 ml. should be sufficient). Heat on the hot plate until all the solid material is In Solution. Pour the Hydrochloric Acid Solution, containing the Iron as Ferric Chloride, into a glazed porcelain evaporating dish (deep type) large enough to contain it and yet small enough to provide as small a surface area of the Solution as possible in the course of the evaporation to follow. (Use a Coors Slayed Porcelain Dish 120 mm. Diameter. Height 50 mm.) *No water* may be used to wash any of the Ferric Chloride Solution adhering to the Beaker into the dish. (See Note 4.) Evaporate the acid Ferric Chloride Solution slowly on the sand bath, care being taken to see that the temperature of the sand bath is not sufficiently high to cause burning or undue drying. *Do not* use the water bath or the steam bath for this evaporation.

As the evaporation proceeds work down into the body of the thickening solution any Ferric Chloride gel which may form and adhere to the evaporating dish around and above the solution level. Use a

17

rubber policeman on a small stirring rod to accomplish this. The solution will finally thicken to a gel, and the evaporation *must not* be carried out to the point at which the gel commences to dry. The gel must remain moist with a slight excess of Hydrochloric Acid.

As soon as bulk of the acid has evaporated and the mass consists of a gel still moist with Hydrochloric Acid, remove from the sand bath and, after thoroughly cooling the dish and contents on cold water and then carefully drying the bottom of the dish with a cloth, making sure that no trace of water is left around the lip of the dish nor has reached the interior of it, transfer the Ferric Chloride Gel to a wide mouthed glass bottle of about 250 ml. capacity, and having a ground glass stopper. Accomplish this by working the gel over the decanting lip of the dish with the rubber policeman. Use *no* water. Pour a few ml. of ether (See Note 2.) into the dish after making sure that no bare flame, nor electric element or sparking equipment is near, and work any gel residue remaining in the dish out with the ether into the glass bottle by means of the rubber policeman on the stirring rod. Ignore any dried rim of Ferric Chloride there may be in the evaporating dish.

Add, *at once*, a 100 ml. portion of ether (have this ready at hand) to the Ferric Chloride gel in the glass bottle and insert the glass stopper. (It is well to have the glass bottle standing in a dish of cold water the level of which is not sufficiently high to tend to float the bottle). Shake the bottle containing the gel and the ether vigorously for a few minutes, cooling the lower half of the bottle frequently under the cold water faucet. Allow to stand a minute, then remove the stopper, first being very careful to dry off thoroughly any drops of water on the bottle and especially around the rim of the stopper orifice, and decant the now brown ethereal solution into an Erlenmeyer Flask of 300 ml. capacity and stopper the Flask with a rubber stopper.

Add another fresh 100 ml. portion of ether to the contents of the glass shaking bottle and repeat the ether extraction by shaking once more for a few minutes, cooling as before under the cold water faucet. Dry the bottle and allow it to remain stoppered. Both the glass bottle and the Erlenmeyer Flask containing the first portion of ether should be kept in a cool place until required or, if this is not available, stand them in shallow running cold water keeping the stoppers dry and protected from water.

Set up a distillation train consisting of a Pyrex distillation flask or retort of 300 ml. capacity, the stopper neck of which is fitted with a bored rubber stopper through which is inserted a 3 inch immersion mercury Thermometer reading from 0 C. to 200 C., so that when the rubber stopper and thermometer are placed in position in the distillation flask or retort the thermometer bulb clears the bottom of the distillation flask or retort by three or four millimeters--a condenser, water jacketed so that cold water may be passed continually through the jacket, filled to the delivery arm of the distillation flask or retort on one end, and to a receiver of sufficient capacity, (about 500 ml.) at the other end.

Having removed the Thermometer and stopper from the distillation flask or retort, arrange a 60 degree Pyrex Filter Funnel (having a diameter at the top of about 2.5 inches) in a ring support or clamp on the retort stand and with the stem lowered into the distillation flask or retort so that the end of the stem is well below the level of the opening into the delivery arm of the flask or retort. Place a fast filter paper in the funnel and fix in place by moistening with a little fresh Ether. Using first the Ether extract contained in the Erlenmeyer Flask, filter this through the filter paper in the funnel. Then replace the filter paper in the funnel with a fresh filter paper and introduce into the funnel the Ether extract contained in the glass bottle together with the

gross body of the Iron, decanting and filtering
first the bulk of the clear brown Ether Solution,
with the last remaining 20 ml. or so, swirling the
gross body of the iron into suspension with the
remaining Ether Solution, decant the whole into the
filter funnel. Allow the ethereal solution to filter
through into the distillation flask or retort, then
wash out the glass bottle with about 10 ml. of fresh
Ether and pour this over the residual gross body of
the Iron in the filter paper. Allow these washings
to pass through into the distillation flask or
retort, then remove the filter funnel and discard
the filter paper and contents.

Moisten the rubber stopper fitted on the
Thermometer, with a drop or two of absolute alcohol
[Ethyl] and insert the rubber stopper and
Thermometer in the distillation flask or retort.
(The alcohol facilitates the proper placement of the
rubber stopper). As a precaution arrange a Burette
clamp on the stand so that one claw of the clamp
bears down, when the clamp arm is screwed in
position on the rod of the stand, on the top of the
rubber stopper carrying the thermometer, and just to
one side of the Thermometer.

Using a Precision Electric Heater equipped with
Rheostat, *not a bare flame*, and arranged so that the
bottom of the distillation flask or retort is about
2 1/2 inches from the heater element, gradually

21

raise the temperature of the contents of the distillation flask or retort until the Ether is distilling at about 40 C., as shown on the thermometer Scale.

Continue the distillation until the volume of the contents of the flask or retort is about 125 ml. Ignore for the time being any suspended solid matter or cloudiness which may have appeared in the solution.

Until the first small portion, about 3 ml., of distillate appears in the Receiver, the Receiver should be left loose at the neck where it fits on to the discharge end of the Condenser. When about 3 ml. have distilled over, the temperature in the distillation flask or retort will have risen to more than 35 C. The Receiver may then be fitted up tightly to the end of the condenser; but until this is done and while the neck of the Receiver is loose, a wet strip of cloth should be wound around the open joint to prevent the escape of any vapor. When the neck of the receiver is closed up with the discharge end of the condenser this cloth may be dispensed with.

Then the contents of the distillation flask or retort approximate 125 ml. in volume, turn off the electric current to the Heater and by loosening the Heater clamp drop the Heater on the stand away from the bottom of the distillation flask or retort and place a sand bath or some such insulation over the element of the Heater to prevent the heat from rising.

Cool the bottom of the distillation flask or retort with a wet cloth. Remove the rubber stopper carrying the Thermometer, and add to the contents of the flask on retort about 100 ml. of absolute Ethyl Alcohol (See Note 3.) Replace the stopper and Thermometer.

Raise the Precision Electric Heater to its original position on the stand and continue the distillation at 45 C. to 50 C., until 40 ml. or 50 ml. have distilled over, and observe the same precautions as before with respect to the Receiver Flask.

At this point any portion of the gross body of the iron which may have passed into solution in the excess of Hydrochloric Acid present with the Ferric Chloride Gel when the Ether Extraction was made, will have been thrown down out of solution (See Note 5), and the excess of Hydrochloric Acid will have combined with the Ether in the first distillation and alcohol in the subsequent distillation to pass over as Ethyl Oxychloride and Ethyl Chloride respectively.

Again disconnect and drop the Electric Heater on the stand. Cool the distillation flask or retort as before and remove from the train. (Note. 4)

Whenever the distillation train is to be broken at any point after heat has been applied to the distillation flask or retort, great care must be exercised for the reason that if pressure within the apparatus has not attained equilibrium with the outside pressure, violent restoration of equilibrium will take place if the difference in pressure is excessive, and this may completely ruin the operation. Always break the train or remove the rubber stopper from the distillation flask or retort only after sufficient cooling has taken place, and then very carefully and by degrees, thus allowing gradual re-adjustment of pressure.

Decant, carefully, the contents of the distillation flask or retort into a Pyrex Beaker of 400 ml. capacity. Wash out the distillation flask or retort thoroughly with a little absolute Ethyl Alcohol, adding these washing to the main portion in the Beaker.

Replace the distillation flask or retort in position in the train and, again using the Pyrex Filter Funnel fitted with a fast filter paper, filter the Solution contained in the Beaker back into the distillation flask or retort, washing the filter paper and contents, after the main body of the solution has passed through the filter, with 5 ml. of absolute Ethyl Alcohol, but first having washed out the Beaker with a like portion of alcohol, and having passed these washings through the filter. When the last washings of the filter paper and contents have passed through into the distillation flask, remove the filter funnel and discard the filter paper and residual contents.

Replace the rubber stopper and Thermometer in the distillation flask or retort, raise the electric heater, so that the heater element is, this time, about one inch from the bottom of the distillation flask or retort. Raise the temperature of the contents of the flask or retort gradually (observing the same precautions as before with respect to the Receiver Flask) until a temperature of about 85 C. is reached, and continue the distillation until the volume of the contents of the distillation flask or retort is about 50 ml.

Disconnect the Electric Heater once more and drop it away from the bottom of the flask or retort. Cover the element with a sand bath as before, to prevent the heat from rising. Cool the flask or retort to about 40 C. Remove the rubber stopper and Thermometer carefully to allow readjustment of pressure. Add 75 ml. of absolute Ethyl Alcohol. Re-insert stopper and Thermometer. Raise the Electric Heater into position, so that one inch separates the bottom of the flask or retort and the heater element, and continue the distillation, observing the same precautions as before, with respect to the Receiver Flask.

Allow the temperature of the contents of the distillation flask or retort to mount a little more rapidly this time until a temperature of 90 C. is reached, and maintain this temperature until the volume of the contents of the flask or retort is about 65 ml. Then allow the temperature to mount gradually until there is a sudden, short and moderately violent ebullition of the Solution in the flask or retort. Allow the temperature to continue to mount until the contents of the flask or retort appear wine red by transmitted light. The volume of the Solution at this point should be about 35 ml. The fluid in the flask or retort is now the true Oil of Iron, of which five, six or seven drops in half a tumbler full of water may be taken two or three times a day before meals with great advantage and without any fear of ill effects, for the general health and especially in cases of anemia and other ailments as stated in Frater Archibald Cockren's great work entitled, "Alchemy Rediscovered and Restored"[1]. The writer wishes here to express his deep gratitude and thanks to Frater Archibald Cockren for the very great amount of help he has received through his study of this book, and to Imperator Ralph M. Lewis and others who made the

---

[1] R.A.M.S. Library of Alchemy, Volume 34.

book available to and brought it to the attention of the membership of the A.M.O.R.C.

Disconnect the Electric Heater and drop it away from the bottom of the distillation flask or retort. Cool the flask or retort and remove it from the train observing the usual precautions with respect to Pressure readjustment. Decant the Oil of Iron into a dropper bottle made preferably of brown glass.

If, at the conclusion of the decanting operation a thin red-brown film of sediment is found adhering to the bottom of the flask or retort, pay no attention to it. It is not caused by the presence of any of the gross body of the Iron. It is due to local overheating of the oil and is caused by there having been too high a heat on the electric heater element. This, however, should be avoided. After decanting the Oil of Iron from the flask or retort, and washing the flask or retort out with a little distilled water (not to be added to the oil) without disturbing this film, and then, upon the addition of a further portion of distilled water, about 25 ml., rubbing off the film with a rubber policeman and agitating it with the water, it will be found that it dissolves in the water to form a bright red solution. Warming the Solution helps to accomplish this.

The Oil of Iron: Notes on the Procedure

Note 1.

The decantation is most easily performed by using the water vacuum pump with a very slight degree of suction and by keeping the end of the rubber suction tube just below the surface of the clear supernatant solution. By watching the progress of the decantation through the wall of the Beaker, any suction of the precipitate through the tube when the solution level is approaching the top of the settled Ferric Hydroxide layer, may be avoided by raising the end of the suction tube and thus discontinuing the operation. This method is much easier and more effective than decantation by manipulation of the Beaker, or "*hand tilting*".

Note 2. ETHER.

Use Ether Squibb (E.R. Squibb & Sons, New York) for anesthesia in copper protected cans, holding 1 lb. net weight of Ether. This may be bought at a very low price from any wholesale Drug Company. I am sure it cannot be made on a small scale in the laboratory as cheaply as it may be bought wholesale.

Before using, rectify the Ether in a glass
vessel (well stoppered) over anhydrous Potassium
Carbonate, C.P. Ten or fifteen grams of Potassium
Carbonate for each pound weight of Ether being
rectified is sufficient. Let stand for 48 hours,
shaking occasionally. *Re-distill just before using*.

Note 3. ETHYL ALCOHOL.

Use the U.S.P. 190 Proof Alcohol Distilled from
Cane Products by the Commercial Solvents
Corporation, Agnew, California, if possible. Redistill at 79 to 80 C., to about 1/15 of the
distillate over Anhydrous Potassium Carbonate, C.P.
using about 25 gm. per liter of alcohol. Let stand
for 48 hours, shaking occasionally. Redistill at 85
C. before using.

NOTE 4. WATER.

All water used should be distilled water. It is
absolutely essential that after the drying of the
Ferric hydroxide precipitate, no slightest trace of
water should be allowed to enter into the procedure,
and every precaution must be taken to prevent it so
doing. Otherwise a proportion of the gross body of
the Iron commensurate with the quantity of water
present and with its solubility therein, will be

carried into the Ether Extract and will not be thrown down in the subsequent procedures, for the reason that Ferric Chloride in solution with Hydrochloric Acid in various degrees dilution with water, is soluble in Ether and subsequently in the alcohol added in the course of the distillations. Thus it will be present in the end product which will not be the true Oil of Iron.

Note 5. THE PORTION OF THE GROSS BODY THROWN DOWN IN THE COURSE OF THE FIRST TWO DISTILLATIONS.

This small portion of the gross body of the Iron carried into the Ether Extract is due to the presence at the end of the evaporation of the Hydrochloric Acid Solution containing the Iron as Ferric Chloride, of the slight but necessary excess of free Hydrochloric Acid, and the Ferric Chloride held in solution by this excess of acid is extracted, as such (Ferric Chloride) by the Ether; the alchemistical Sulphur and Mercury of this portion of the Iron being unable to separate from the gross body to form the coalition with Ether as the Ethereal Oil.

Upon distillation, first of the Ether extract and later upon the addition of alcohol, the Ether Solvency factor with respect to this portion of Ferric Chloride is destroyed and coincidental separation of the gross body of this retained portion from its five parts takes place, the gross body constituent being thrown down out of solution and the alchemistical Sulphur and Mercury constituents, forming their proper coalition, are retained in solution.

May we not, with respect to this small portion of the gross body of the Iron which we are able to separate and leave behind only after this necessary further processing, perceive a parallel in the statement made by our beloved Brother, Dr. H. Spencer Lewis, late Imperator of A.M.O.R.C. to the effect that most of us like to carry with us some of our yellow flowers along the Path of Initiation only to find that we have to drop them and leave them behind at some point in our journey.

OIL OF VITRIOL

From Flints, Iron Pyrite, Magnetite. Iron dissolved in Aqua Fortis: and alcohol distilled together. Then add True Oil of Tartar.

* * * * * * * * *

ANOTHER

Grind Magnetite to light powder (or Pyrite). Dissolve in HNO_3, Dilute slightly. Heats and foams in time. Remove Poison Brown bases. Filter to Brown/Red liquid. Use Ammonium Carbonate powder to neutralize. Iron Hydroxide $FeOH_3$ will precipitate. Takes Time. Decant filter and Wash. Dry powder to red Brown.

Digest 1 month with acetone menstruum (sharpened with Ammonium Chloride NH_4Cl to orange.) Distil to clear first. Then extract with clear menstruum in Soxhlet, color changes 1 day to 1 week. Filter and gently distill (Roto Vapor) oil does not distill – it is immature Oil of Iron.

Mature by digesting with Iron Hydroxide or by circulation. Gains more color and oiliness. Filter. Distill Volatile part with Vacuum leaving red oil. Oil dissolves in alcohol which purifies it. 200 C. in Vacuum will leave Gum and Separate light red oil, Gum will produce more oil, with alcohol digestion.

Oil will be cleaned and sweetened with alcohol digestion.

* * * * * * * * * *

The ALKAHEST is a Mineral Substance or FROM a Mineral.

The PHILOSOPHICAL MERCURY is from a Metal.

The Alkahest dissolves the Sulphur; The Philosophical Mercury dissolves the Body and Essence.

RECIPES

**Being various notes from such sources
as the Paracelsus Research Society,
Heydon's "The Rosicrucian Crown" and
from the writings of George Ripley.
These notes are various recipes
utilizing different metals such as
Iron, Lead and Antimony.
R.A.M.S. 1985**

WHITE WIFE.
PROCESSES ON LEAD.

A Viscous water extracted from Jupiter's Bowels
i.e., White Lead=lead Acetate is moist and wets the
finger. It is a Coagulated Mercury not yet fixed.
Spirituous and Volatile.

Within a month should be put on its Calx. Will
boil by itself (sealed shut), until dried up in the
Calx. After Conjunction it looks like common Mercury
and does not wet the finger.

GALENA-PbS.

Using dry ore and extract white & red Mercury -
No Menstruum.

This is LIVING LEAD PbS Ore. Roast it to turn
it to oxide. Then using acetic acid from Red Wine

Vinegar make Acetate. Acetic Acid opens up all the metals.

TO PRODUCE SUGAR OF LEAD.

Place powdered Ceruse into Matrass pour on 12 to 15 times as much acetic acid (strong vinegar). Place in Sand Bath 1 day, to Digest and shake time to time. Decant liquor, put on fresh Vinegar and digest again. Proceed until Ceruse is 2/3 gone.

Evaporate to a pellicule the liquor and place in cool place. Dry crystals will shoot up. Decant and evaporate till crystallization stops. Redissolve and re-evaporate to purify.

LEAD ACETATE - SUGAR OF LEAD

Heat in Crucible to Red hot. Distill in reverberatory; an austere liquor.

Distill off about 1/2 of this. An inflammable Spirit.

A yellow Oil will come to the top, a blood Red Oil will go the bottom. The Spirit remains after oils are removed and has a very fragrant odour.

Distill the Acetate & Calcine Black dregs to RED Pb_2O_4, make acetate again.

The Acetate is the SALT of Lead.

The Special Sublimation Cockren talks about is the dry distillation of the acetate (after washing with Vinegar of Antimony).

The Acetate is simply said to be washed in the Vinegar of Antimony before Distillation or Sublimation as Cockren calls it.

The Oil is not fixed. Only the Spirit is.

The Spirit (Acetone) is not fixed but the Vinegar is and Fixes what it Contacts like Regular Vinegar. This is the Key that most seem to be missing for Mercury.

The Destructive distillation of the acetate yields an acetone & an oil which must be separated to the spirit and oil, i.e., Fractional distillation (Dexterous Distillation).

FURTHER METHOD.

Wash acetate with KM (In one old *Parachemy*[2] the KM is said to be Vinegar of Antimony in one of Frater's quotes.) Then distill off KM. Then a white clean water comes over. Then an oil which is only Part of the Sulphur. Circulate this with KM=unfixed oil.

---

[2] *Parachemy* was a periodical published by Frater Albertus Spagyricus (Dr. Albert Richard Riedel) from 1973-1979.

FURTHER PROCESS.

Putrefy 42 days then distill lead over at 22.5 C. Repeat Process.

It will then dissolve a little gold to Golden Yellow white. A Hydrocarbon Compound?

Second step in Philosophical Mercury is Purification. Third Step is Cohobation. The Gold used is Gold Chloride (said in one of Frater's Interviews) for producing oil of Gold - not the Stone itself undoubtedly.

The Stone of Gold is said to be in Lead by Paracelsus. He says this is A WATER with which the Spirits of the six metals are congealed into the essence of the Seventh.

A MORE COMPLETE PROCESS.

Reduce to Acetic Salt and distill; condense to Golden Water (from body of Antimony) Dry ice in acetone both can be used. Separate this to a white Mercurial water=Primary water, does not wet the finger.

A red tincture can also be obtained which deepens in time.

If white and Red portions of Golden water are joined they form a deep amber liquid. Shows light from Sun intensely. Purify dregs-calcine to redness and treat to the white Salt (acetic). Join in exact quantity and seal flask.

Exact heat is required. After heating forms leaden mud and grows crystals which change in colour. Raise heat. It all melts into amber fluid.

This becomes like black pitch; add more ferment or mercury; repeat heat and redissolve. Heat to colours - then white, then Citrine, then red.

FURTHER PROCESS FOR ACETONE - P. MERCURY.

1st. dry distillation of acetone yields yellow residue. Retreat this with Vinegar and repeat.

Digest distillate 14 days and distill. First flood=Spiritus Ardens. Rectify this - white oil appears on surface. Yellow oil remains behind.

Take sublimate from neck of retort, grind and place on cold iron to dissolve. Filter and add a little acetone, A green oil will appear on the surface, distill, first water then oil.

Redistill water and evaporate in Water Bath to thick oil. Treat with a little acetone, Gold is dissolved by the oil. This is acetone Aerrimum or *Dissaeveus Auri*.

Distill Spiritus Philosophii over Sal Tartarii 3 times.

Digest 50 days - yellow residue on bottom.

3 fingers of this on Gold and Silver separately after resolving these in mercury and Fuming away; w/Bath first, then ash both at boiling temp.

Decant yellow/gold water; decant green and blue/silver water; repeat until all is dissolved; digest 40 days result.

Distill out Solvent leaving oil of the Metal. Pour back distillate; apply W/Bath 24 hours.

Distill gently in Sand. First water, then higher temp. for Spirit. Higher still for part of oil.

Add water to distillate. Digest and Distill in Sand. Repeat until all passes over. Rectify 7 times each. Mix both Silver and Gold and Circulate 60 days.

THE DRY WAY WITH THE GUM OF ANTIMONY.

(The Gum is the Yellow Powder remaining after Acetic Vinegar Extraction of Antimony Glass). Dry distillation at this point is the DRY WAY.

Basilius used the WET WAY.

The Dexterous Distillation Frater A.[3] discussed (when separating an oil from feces) was for either way, Wet or Dry. Very difficult the Dry Way though.

DEXTEROUS DISTILLATION of the oil over the helm is FRACTIONAL DISTILLATION.

The Wet Way: the Spirit of Antimony can be used.

CERUSE = Flemish Almalis (an oxide)

MINIUM = Red Lead (Mercury of Saturn Precipitated)

LITHARGE = Yellow oxide of Lead.

WHITE LEAD = Lead Carbonate. Calcine Lead Carbonate to get Minium.

In acid Solution lead can be precipitated by Potassium Carbonate into Lead Carbonate.

GEORGE RIPLEY.

Sericon or Antimony.

Sericon = An Iron Compound (Pliny).

Sericon = Red lead = Minium (Ruland).

---

[3] Frater Albertus

RAW ORE.

Dissolve each 1 lb. ln 1 gal. of Distilled Vinegar (2 times distilled) evaporate the Vinegar to a Green Gum or Crystals (use low fire) dry distill gum-yields White Smoke. Smoke turns Reddish, change Receiver and use Dexterous distillation for Red Oil to Ascend. (From a talk with Frater Albertus).

DESCRIPTION OF LEAD PROCESS;

AND DISCUSSION W/FRATER A. BY STUDENT.

Take GALENA - grind; dissolve in 1/3 Nitric Acid 6 N., heat gently (Brown Fumes are dangerous: Hydrogen Sulphide).

Wash out Nitrate or Neutralize with 6 N., Potassium Carbonate Solution.

WASH Carbonate to white powder. Calcine to Red Lead Oxide. And 6 N., Acetic Acid (1/3) Red Lead dissolves - Carbonate Bubbles. Acid Receives green Tint. Evaporate with digestive Heat to Gum or Paste (Crystals).

Dry distill these Crystals to achieve Spirit. The Philosophical Vinegar can be used to dissolve & reduce Red lead to a Black Mass.

After Green Lion is treated - it expands and Puffs up. This is Black Dragon During distillation.

The Flood comes over first. =Phlegm.

Then a Golden Coloured Liquid comes over smelling of Acetone. If mass is still wet it will ignite like Phosphorus in air. (If dry - place in glowing Coal and Calcine it to Citrine.)

The WET WAY usually produces the most violent Bursts. The Dry Way is Safer.

In Acetone Extract of the Dried Kermes (from lye dissolution) use Hot Soxhlet. Will turn Green in a few days. Then blue, then yellow. WITH THE RED OIL=the 4 Elements. Remove Acetone and Circulate w/KM.

In Cohobating a mixed Fluid, a Sealed Glass and a Steam Bath are used.

SEPARATION and PURIFICATION of oils can be done by a thread through a straw by osmotic Pressure. It must pass over the Helm.

PUTREFACTIONS IN THE WORK (FULCANELLI). After refining oils they can be Fermented.

1st. Separation shows some Acidity and smells of the Grave-Foams and shows Fish Eyes.

All Putrefactions turn BLACK.

2nd. is the first Conjunction,

3rd. Second Conjunction of Heavy Water & Sal.

4th. Fixation of the Sulphur.

ANTIMONY & THE FIRE STONE.

Antimony is first to be prepared into a true Stone-The Quintessence. It performs what Aurum Potable does.

Mercury of Antimony is matter to produce Arcanum of the Stone called Heaven of the metals - a pure Quintessence.

With the Fixed Sulphur of Vitriol in the Magnetic Stone one can ripen the Mercury or Heaven.

Antimony has the same essential root as Mercury; Antimony has more salt though. Vulgar Mercury has no Salt, but has very hot Spirit of Sulphur.

The Spirit of Saturn can bring it to Coagulation and Fixation.

Vulgar Mercury is mere fire which resolves itself into an incombustible oil spiritually

The Black Becomes:

Blue=Earth

Green=Water

Yellow=Air

Red=Fire

This may be Philosophical Acetone action on Stibnite dissolved in Lye (Solution) and precipitated with 6 N. Vinegar & dried-use Soxhlet (hot). It passes through colours:

1st. Green

2nd. Blue

3rd. Yellow=Gold.

Antimony has the same operation as Corporeal Gold, has descent from the same ORIGINAL as the Star of Mercury. The Mercury of Antimony is to be separated from its SULPHUR by a natural Method. (Basilius says elsewhere that "Mercury of Antimony is in the Regulus", but this cannot be the pure metal at all but SULPHUR here must also be the "Real Regulus" The Son.)

ANTIMONY.

The Four Elements are:

Air- Spirit or Mercury

Earth- Mercurial Salt

Fire- Sulphur

Water- Radical Moisture

The radical moisture separates by distillation from the Spirit.

The "Tincture" or "Antimonial Sulphur" is equivalent to Potable Gold. This purges the Soul of Gold while Gold ameliorates the Soul of Antimony to Fixation, exalting both.

The Spirit Penetrates

The Body Fixes

Soul joins together, tinges and whitens.

These three made subtle and volatile to a Golden Coloured Water are The PHILOSOPHERS MERCURY.

(The Green Gum or Green Lion) from yellow glass of Antimony make a fixed tincture. This gives white and yellow feces.

Dry digest these to Ice Blue Crystal (feces) giving a GREEN Extract=Gum or Lion.

ANTIMONIAL QUICKSILVER.

Mercurial Water, heavy, Viscous and Glutinous. Dissolvent for oil and water or Spirit and Soul of the Sun and Moon.

Antimonial Quicksilver=Secret Fire, Piercing Vinegar, Living Water, White Smoke, May Dew.

The whole work is to reduce Gold and Silver to a precious Oil with this water. This oil is then the Quicksilver of Gold and Silver.

ANTIMONY.

From it Aurum Potable can be made.

Spirit of Mercury. The three essentials of antimony can be brought together by "Archeus" of Element Earth by distillation from antimony; Proceedure:

Vinegar/acid

Red colour/sweet

Bitterness

Salt oil/acrimony

To rid Sb_2S_3 of Sulphuric acid one way, distill w/Rain water 1 month, then Calcine.

RUNNING MERCURY OF ANTIMONY.

(Further process.)

Take 8 parts of Regulus

1 part Salt of Urine (clarified & sublimed)

1 part Sal ammoniac

1 part Salt of Tartar

Mix all salts and pour on Wine Vinegar. Digest 1 month. Place in ashes & distill off Vinegar; dry Salts

Mix with 3 parts Borax (Venetian Earth) Distill by Retort in Strong Fire; Putrefy this spirit with Regulus reduced to powder. (2 months) Distill off

46

Vinegar gently, mix feces with Iron filings, 4 parts; Distill Violently.

Spirit of Salt carries over the Mercury of Antimony as a fume. Use a large recipient for this with water contained so the spirits of salt mix with the water but the running Mercury goes to the bottom. Living Mercury; Take this Mercury expressed through a skin, 1 part.

Pour on 4 parts highly rectified Red oil of Vitriol (True Oil of Iron). Extract oil and Spirit stays with Mercury. Increase fire to sublime. Put down sublimate. Repeat 2 more times with as much oil 4th. time grind with all Sublimate & Feces (Earth) to clean Crystal.

Place in Circulatory and pour on a like weight of oil of Vitriol and 3 times as much S.V.

Circulate till separation is made. The Mercury resolves itself into oil, Floats like olive oil-separate oil. Circulate with strong Vinegar 20 days. Oil settles to Bottom. Venom remains in Vinegar. If you can bring this oil to a fixed Stone it is then a fixed Tincture for Metals.

OIL OF MERCURY.

Running Mercury 1 part (antimony) Red oil of Vitriol 4 parts. Digest together-Some time; Extract oil and

Spirits of oil stay with Mercury. Sublime. Repeat all 4 times to crystal like mass; add equal weight of oil of Vitriol and 3 times as much S.V.; Circulate until the Mercury resolves itself into oil. Floats like olive oil. Separate. Add Strong actuated Vinegar and digest 20 days. Oil Settles.

PHILOSOPHICAL MERCURY

There are Two Kinds:

Male & Female

One is Eastern=Soul

One is Western=Spirit=Tincture

The Tincture or Spirit is used to Transform the Soul.

The Tincture is fixed to the Stone.

TINCTURE=Water=Fire of the Stone.

KM=Alkahest is distilled Vinegar of Antimony, residue is no Good.

(Frater A.) Mercurius Philosophorum is Vinegar of Antimony (Prepared and Elaborated). Take Volatile Mercury, distill from body. Coagulate in White Salt or Philosophic Sal Ammoniac (also from Kuhul = antimony ore).

VINEGAR PROCESS.

(Frater said that normal Vinegar "opens" all the metals producing Acetate. He also said in a Parachemy the exact same thing for Vinegar of Antimony, it "opens" the metals. He also said both fix what they come in contact with.)

Antimony (Stibnite). Calcine gently in Vacuum to remove Sulphur without creating acid. Then it is ready to make Vinegar.

Vinegar of Antimony is the 1st. Key to the 1st. Gate and the Last (The Kings Palace, i.e., The Ferment). This Key dissolves, opens, divides and separates.

To vivify Antimonies Mercury it must be precipitated to a fixed powder. Then an oil can be made.

VINEGAR (Article from Essentia[4]).

Is NOT Sulphurous Acid or Sulphuric Acid; NOT Hydrogen Sulphide. Refuses to combine with Alkaline Salts.

It IS volatile and Acidic. Does not precipitate Metallic Salts; no effect on organic reagents. IS a Highly Volatile liquid. Boils at room temperature; composed of (Sulphur, Hydrogen and Oxygen).

---

[4] Essentia was a periodical published by Frater Albertus Spagyricus (Dr. Albert Richard Riedel) from 1980-1984.

VINEGAR OF ANTIMONY.

It cannot be analyzed properly. It forms no well-defined Compounds.

6 of ore. 14 to 15 of distilled water (another says 6 of ore to 16 of water) Long neck phial. Dung heat 40 C. Water Bath. Foam is Hydrogen Sulphide Gas. Distill with beak in Receiver Water; don't distill to dryness. Sublimate for 3 days total (Calcine) 200 C.

Sublimate on head is yellow to red.

Grind sublimate into Feces and add back water; redistill immediately, do not allow to digest any length of time (becomes weaker). Repeat Process adding fresh ore 3 times. Becomes very Acid. Produces 3 things in Sublimation;

1st. Red allotrope of Sulphide

2nd. Black Sulphide

3rd. $Sb_2(OH)_2S_2$= Hydroxy Sulphide. Fumes will cause headache.

Vinegar of Antimony (Adopted from Basilius' Triumphal Chariot of Antimony[5])

On Fermenting Corn.

Corruption of the ore is by steeping ln water, draw off water, digest in heaps, dry and separate fine. (Coagulation-Reverberation). Grind small (Calcination). Imbibe water back again. Digest and distill. Add Its Salt to ACTUATE. Then a ferment is added to excite internal heat. Sublimate the Spirit to purify it. In Antimony and Mercury all colours can be found.

Vinegar of Antimony. Powder ore, Pour on Vinegar and expose to Solar Rays (heat) to red. (40 days). Decant and gently distill. Extract Gum/Powder with Spirit of Wine (Maybe Philosophic Mercury); to a red Tincture. Circulate to Volatilize. Distill to red spirit and abstract spirit to a thick red oil (dexterous). This oil must be united to its salt for working with metals. The Vinegar also is united to its Salt and distilled strongly in ashes for True Vinegar.

ACTUATING VINEGAR WITH SALT.

The first kind of actuation is done by the raw ore digested with 3 fingers of Vinegar. Takes a red

---

[5] Volume 2 in The R.A.M.S. Library of Alchemy.

Tincture. Distill Vinegar-Red stays behind. Extract red with S.V., for Medicine.

Rectify Vinegar and dissolve its Proper Salt.

iij of Vinegar

j of Salt.

Force strongly by ashes, Vinegar gains Sharpness.

Vinegar of Antimony.

Actuate with its own Salt first. Extract raw ore (probably calcined) 40 days, filter and putrefy another 40 days to a black colour; (True Solution) from which separate the Elements.

Distill- grind remaining matter- Distil with water. Dry.

Sublimate 2 months with S.V. (Philo. Merc.). Moderate heat. Remove Feces. Distill in ashes; Turns Golden when Elements separate. Gold matter subsides- Red oil goes over. Circulate Red oil 10 days, oil separates out of this (is Sweet); Vinegar needs PUTREFACTION and re-distillation to be pure Fixed Spirit of Antimony.

BLACK DRAGON.

(Frater A.) Is black dregs of Antimony work which yields Dragons Blood or Blood of the RED LION. After Earth yields oil then Calcine Violently and clarify w/"vinegar" to make TRUE SALT.

(Frater A. & Basil) Antimony is Frigid and Humid: Hot and Dry: Volatile and Fixed: its Volatility is poisonous: Its Fixed State is free of poison.

(Frater A.) Antimony has the Alkahest within it.

(Frater A.) The Opus Major is just an incident in the Essential Process or Opus Minor.

(Frater A.) Antimoneous Acid is Vinegar of Antimony but contains no Carbon like the plant Variety. No other metal or mineral contains a FIXED ALKAHEST.

The Philosophical Mercury is one of SEVEN Alkahests.

Low Boiling points are indicative of more penetrative ability into Minerals.

Frater says any Salt of Antimony or Glass can be extracted by Antimoneous Acid or: Vinegar of Antimony.

He then mentioned that Antimony Pentoxide (heavenly calcined) had a similarity or relation to said extraction.

This Vinegar is fixed by nature not by man.

The Mercury or Essence is the fixed Spirit of Antimony. The Quintessence is the oil (red).

Take Antimony ore pounded fine. Pour on distilled Rain Water which has been electrically charged. So it contains activated Nitrogen. This will then effervesce. DUNG HEAT. Some of the Gang provides H_2SO_4, during reaction.

Distill off water. Change Receiver. Dry Distill Vinegar over as a White Smoke. Repeat Procedure. "Vinegar of the Sages". Add a little water to it to extract glass or oxide to a red Tincture.

(Frater A.) Copper, Silver and Mercury are the normal subjects for Transmutation. He stated that the Philosophers Stone is made from Gold.

SULPHUR OF ANTIMONY.

Take Crude Antimony ground and Vinegar (actuated) with Salt. Putrefy (B/M) 40 days Decant red tincture and add more vinegar till there is no red tinge. (Can use Soxhlet). Place all tinctures together and putrefy B/M 40 days. Matter becomes Black as INK. This is true solution. Distill clear Vinegar and grind the feces. Sweeten with rainwater and dry gently.

Place in long Circulatory with 3 fingers of S.V. Moderate heat for 2 months. Filter wine and

54

distil in ashes, till S.V. carries over Red oil, separate Black dregs after S.V. maceration, when distilling oil & S.V. a Golden Sheen is in vessel. Red oil passes over. Circulate this 10 days. Oil separates out to BOTTOM. S.V. on top. Separate in a separatory. Very sweet. Balsom of life- Sulphur of Antimony. The Dose of this Sulphur or Quintessence is 8 grains before fixation.

Does all that Aurum Potable Can.

The Star of Gold and The Star of Mercury arise from the same blood of their Mother.

After Coagulating this Sulphur. This Sulphur is used for the Fire Stone.

Dew changes and matures all Salts Because of its subtle Nitre.

The Mercury and Sulphur are added separately to the Salt of a Metal.

1 part Salt to 1/8 part Simplex (Vinegar) dry gently 3 or 4 days- Repeat; don't use too much moisture. Repeat 8 times to make equal weight

Should flow like wax. Then Ferment with Silver Filings to White Elixir. Continued imbibitions increase virtue (before ferment).

Salts are Keys for the preparation of Stones.

There are Salts for opening= Mineral Salts.

And Salts which fix and enter the Product itself= Metallic Salts.

Ferment the body to yield Celestial entity. The lesser fire of Microcosm reduces again to fixed body.

The Stone is penetrating and fiery and is cocted and maturated by Fire.

The Fire of the Elements is Sun and (Ripens).

The fire of flames- corporeal fire (cocts).

For 16 (6 of water

use 6 (6 of Antimony,

Keep the neck in water when distilling.

SALT OF ANTIMONY.

Make Regulus by Tartar and Sal Nitre- Grind subtly place in large round Glass in Sand heat to sublime. Keep pushing down highest powder till all remains fixed on bottom. This yields pure Red fixed precipitate. Regulus = Takes much time. Grind this subtly and place on marble in cool place to dissolve. Takes 6 months. A Red and pure liquor forms. Separate feces.

The Salt became the red liquor. Filter this and extract phlegm gently. Again place in moist place. It will yield Reddish Crystals. Repeat until WHITE.

This Salt actuates all Menstruums for the extraction of Metals.

Dry white salt and mix with Borax 3 parts. Distill this.

First a white Spirit, next a red Spirit which resolves itself into white. Rectify this Spirit in dry or moist Balneo (subtly). This White oil from the Salt of Antimony purifies whole Blood like Salt of Gold (Spirit of Salt of Antimony). When Salts absorb moisture from the air they also absorb free nitrogen which enriches the distillate therefrom.

THE FIRE STONE

Living Mercury of Antimony, Red Oil of Vitriol made of Iron, Spirit of Salt and Antimony.

Mercury is in the Regulus; Sulphur is in the Red Colour; Salt is in the Black Earth.

To separate these and join them again and fix them is to prepare the Stone of Fire.

The Basic Work is the extraction of a tincture from the minera of Antimony BEFORE making glass.

Stone of Fire will tinge 5 parts of Silver, Tin or Lead.

Fixed powders that tinge are Stones.

1st. is Stone of Philosophers.

2nd. Stone (Tincture) of Gold and Silver.

3rd. Tincture of Vitriol.

4th. Tincture of Mars.

5th. Tincture of Jupiter and Saturn.

6th. Tincture of Mercury.

Venus and Mars contain tincture of Gold in themselves when reduced to fixation. Jupiter and Saturn are for the Coagulation of Mercury.

Virtues of All Things, Mineral, Animal and Vegetables are in Stone of Philosophers.

The Oil of Antimony used for the Stone is sweet and casts beams like a ruby in sunlight.

THE STONE.

Make the Sulphur of Antimony; purify & etc. The Sweet Oil.

Mix iiij Salt of, Antimony w/16 ij extracted oil. (2 lbs. of oil, with 4 oz. of Salt).

Circulate one Month-well closed- separate Mercury B.M. then distill off pellucid red oil from feces. (many coloured).

Rectify in B/M until 1/4 can be distilled.

Take result of Running Mercury process. Take equal parts of this crystalline substance and the pellucid oil.

It is the oil of the iron Process, the precipitate from Mercury of Antimony Process.

Of Antimony from earlier process, Seal in phial and they will join and become fixed to each other. (Gentle Heat). The phlegm will be consumed and what remains will become Red, dry, fixed and fluid and will give no smoke or fumes. (A Powder). This will take many months.

They will become a Hermaphroditic body.

THE REGULUS THE SON.

2 or 3 grains in white wine, expels evil. The Spirit from the Salt Process is used here as well. (with Red Oil).

This Regulus is the Regulus which contains the Mercury of Antimony mentioned previously. I believe!

The Third Book.

Of Saturn or Lead the first Direction.

Chapter 1.

Of the Elixir, Putrefaction into Sulphur, the Oil of Sulphur, of the Conjunction of the Salt and Oil of the Spirit, or Salt of Saturne, which containeth the Oil or soul of the Menstruum of White Mercury, and red water of Paradise, Resolution, Solution, distillation, Hyle, Purgation, resolution of Sericon, of the Gum of Sericon, of the solution of thye Minium or Adrop, of Calcination of Minium into Adrop and red Lead, of Calcination of Lead withy AQUA FORTIS.

Very many have writ of SATURNE or Lead, but none that I know of have writ fully thereof in any particular Treatise; therefore I do not here only set down what I have gathered from them most briefly and truly, but also those things which I have found and proved by my own experience, which I have annexed to them, that the work may be absolute and compleate.

Of which, as they say, MARY the Prophetress, and the Sister of MOSES in her Books of the work of SATURNE is thus said to write. Make your water running like the water of the two Zaibeth, and fix it upon the heart of, SATURN: And in another place, Marry the Gum with the true Matrimonial Gum, and you shall make it like running water. Of which process of MARY, GEORGE RIPLEY our Country man hath these verses.

> Maria mira sonat
>
> Quae nobis talia donat
>
> Gummis cum binis
>
> Fugitivum fugit inimis
>
> Horis in trints
>
> Tria vinelat fortia finis
>
> Fila Plutonis
>
> Consortia jungit Amoris.

Or thus,

> Maria mira sonat, breviter qui talia donat
>
> Gummi cum binis fugitivum fugit in imis
>
> Horis in trints tria vinclat fortia finis.
>
> Maria Lux roris ligam figat in tribus horis
>
> Filia Plutonis consortia jungit Amoris
>
> Gandet inassala sola per tria sociata.

The heart of Saturn, saith RIPLEY, is his white and clear body, out of whose doctrine the work doth briefly thus proceed, that is to say, that a water

he made out of the body of Saturne, like the water
Zaibeth, and that water fixed upon the heart of
Saturne; but because the practice of drawing out
this water of Zaibeth, doth not appear out of this,
nor the way of making the heart of Saturne,
therefore the foregoing direction in the HOLY GUIDE[6]
will shew them both.

Therefore I have joined two Tables, in one of
which the shorter is the demonstration of the
reduction of the body of Saturne into his heart or
Salt, the other longer and greater, is the
extraction of the water Zaibeth, and the
consummation of the work of Saturn.

Having thus described this work, I now come to
the explanation, and say, that the Calcination of
the Body is twofold; for the Calcination thereof in
the shorter work, for extracting the heart of
SATURN, is done on this wise by AQUA FORTIS.

Take 8 or 10 Ounces of Lead in filings, and
dissolve it in AQUA FORTIS in double proportion, and
fortified with Salt Armoniack in an Earthen Vessel
with a narrow neck, and set it in ashes till it be
totally dissolved; and there will remain a white
matter in the bottom like Grains of white Salt,
which is a figure of perfect solution; then pour

---

[6] The R.A.M.S. Library of Alchemy, Vol. 40 and Vol. 41.

your matter that is dissolved in the water into a body, and set thereon a Limbeck, and in Balneo draw away the corrosive water, till there remain a dry substance in the bottom; and so you have the body converted white by Calcination with corrosive water, out of which the heart of SATURNE is to be drawn.

The way to wash away and purge the corrosive water from the body, pour warm water upon the substance in a Limbeck, and pour it often off till it have no sharpness at all upon the tongue, and then your body is prepared for drawing out the Salt.

When your matter is well dried, dissolve it again in distilled Vinegar, and distill the Vinegar twice or thrice from it, and in the bottom you shall have a lucid clear and white shining Salt, which is then called the heart of SATURN.

Now I come to the practice of the other greater work, that the verity of the stone may be found, of which many have made mention in their Books, as RAYMUNDUS, who called it the Vegetable, Mineral, and Animal Stone; GEBER saith there groweth a Saturnian Herb on the top of a Hill or Mountain, whose blood if it be extracted, cureth all infirmities.

RIPLEY writ a whole Book, called his PRACTICAL COMPENDIUM, of the practice of the Vegetable Stone, teaching the manner and form of operation; but because he neither set down the solution plainly nor perfectly, he hath been the cause of much error, and hath not only deceived me but all those that followed him, until after a long time I found a way to dissolve SATURN, so that it could never after by distillation be turned into Lead again, which is the chiefest and greatest secret of the Vegetable Stone.

But let us hear the words of MARY the Prophetess, and RIPLEY taken from her: The Radix of our matter is a clear and white body which putrifieth not, but congealeth MERCURY or Quicksilver, with its odour makes its water like the running water of the two ZAIBETH (alias ZUBECH) and fix it upon the fixed heart of SATURN; which words do most aptly agree with the properties of Lead; for if anyone be smit or wounded with a Bullet, and the Bullet remain in the body, it will never putrefy.

And also if Quicksilver be hanged in a Pot over the fume of molten Lead, so as the fume of the Lead touch the Quicksilver, it will congeal it.

Thus far of the preparation of Lead, we now come to its denomination. They bid us fix the water ZAIBETH upon the fixed body Of the heart of SATURN; now for the exposition of the body, for the name of SATURN, RIPLEY calleth it ADROP, of which that is made which the Masters call SERICON; the water of SERICON they call their Menstruum, the two ZAIBETHS joined together in one water, are the two MERCURIES, that is to say white and red contained in one Menstruum, that is to say of the water and Oil of the fixed body or heart of SATURN: Follow what I have written concerning the imbibition of the earth, our operation is no otherwise then in the Practical Compendium of RIPLEY.

ISAACUS also writ a Treatise of Lead, he worketh chiefly according to the doctrine of MARY the Prophetress, and laboreth much to fix the earth of SATURN, and after to dissolve the body in distilled Vinegar; that by the addition of corroding and sharp things, his red Oil may be distilled, which he called the water of Paradise, that he may imbibe his fixed earth therewith; which way is much shorter than RIPLEY'S, but the rubification and fixation of the earth is long and uncertain; wherefore I have both forsaken ISAACUS and RIPLEY ln making the earth, instead of which I have given the fixed heart of SATURN, as you may read in the HOLY GUIDE.

But that the body may be prepared according to this Table, and after my intention and the desire of RIPLEY, we both will that the Oil or Water of Paradise be drawn out of the Gum of SERICON (whose father is ADROP). SERICON is made of Red-lead; therefore it is first necessary to shew the way of making Minium of Lead, which THOMAS JUC[7], an ENGLISHMAN, hath described, together with the Composition of the Gum of SERICON, which AUTHOR I propose to follow, as being the best.

Take ten or twelve pound of Lead, and melt it in a great Iron vessel, as Plumbers use to do, and when it is molten, stir it still with an Iron Spatula till the Lead be turned to powder, which powder will be of a green colour; when you see it thus, take it from the fire and let it cool, and grind that powder upon a Marble till it be impalpable, moistening the powder with a little common Vinegar, till it be like thick honey, which put into a broad Earthen vessel, and set it on a Trivet over a lent fire, to vapour away the Vinegar and drie the powder, and it will be of a yellow colour; grind it again and do as before, till, the powder be so Red as Red-lead, which is called ADROP; And thus is SATURN calcined into Red-lead or Minium.

Take a pound of this Red-lead and dissolve it in Gallon of Vinegar, and stir it with a stick three or four times a day, and so let it stand in a cold place the space of three days; then take your Earthen Vessel and set it in Balneo twenty four hours, then let it cool and filter the liquor three

---

[7] This name is exactly as it appears in the original R.A.M.S. copy. I have been unable to identity Thomas Juc. -pnw

times; and when it is clear, put it in a body with a Limbeck thereupon, and distill the Vinegar so long as it will ascend, and in the bottom the Gum of the SERICON will remain like thick honey, which set apart, and dissolve more new Lead as before for more GUM, till you have ten or twelve pound thereof.

Now give careful attention, for we now come to the point and period of RIPLEY'S error, for if you put four pound of this SERICON to distill in a Limbeck, and from thence would draw a Menstruum, as RIPLEY teacheth, perhaps you would have scarce one ounce of this Oil, and some part of a black earth will remain in the bottom, and most part of the Gum melted again into Lead, by which you may know that the SERICON is not well dissolved, nor as yet sufficiently prepared, that a Chaos may be made thereof fit for distillation, because it is not yet well dissolved; therefore in ISAACUS there is found a way of resolving this Gum with distilled Vinegar, actuated with calcined Tartar and Salt-armoniack; Wherefore, saith he, if thou be wise, resolve thy Gum; but I like not this actuation of the Vinegar, as I may call it. I rather choose to resolve the SERICON in RAYMUND'S calcinative water, which is a compounded water of the Vegetable MERCURY or fire natural, with the fire against nature, as RIPLEY testifieth, and it is more verified by RAYMUND in his Book of MERCURIIS, where he teacheth how to dissolve bodies with his calcinative water.

I will reveal unto you this water, which is almost unknown: Note therefore, that the Vegetable MERCURY is the spirit of Wine (instead of which we may sometimes use distilled Vinegar) and that the fire against Nature is a corrosive water made of Vitriol and Salt-peter.

Therefore take which you will, either spirit of Wine rectified (or AQUA VITAE) or distilled Vinegar four pound, and two pound of corrosive water, and mix them together.

In this water thus compounded, resolve half a pound of Gum of SERICON in a circulatory, and set it in Balneo four or five days, and the Gum will be totally dissolved into the form of water or Oil of a duskish red colour.

Then distill away the water in Balneo, and there will remain an Oil in the bottom, which is then the Chaos, out of which you may draw a Menstruum containing two elements; and this is the true resolution of the Gum of SERICON, in this water you may resolve so much Gum as you please by reiteration.

Take two pound of this Chahodical substance, and prepare it for distillation in naked fire or sand, and lift up the clear red Oil, wherein

both the spirit and soul doth secretly lie hid, which ISAACUS calleth the water of Paradise, which when you have you may rejoice, for you have gone through all the gross work, and come to the Philosophical work.

Therefore now proceed to conjunction, and join the white heart of SATURN with the red Oil, as it is found in the Rosary.

CANDIDA SUCCINCTO JACET UXOR NUPTA MARITO, That is to say, the red MERCURY to the Salt; if you proceed to the red work.

Therefore take four ounces of the Salt or heart of SATURN, and as much of the red Oil or water of Paradise, and seal them up in a Philosophers Egg, and so soon as they shall feel the heat of the Balneum, the Salt will dissolve and be made all one with the Oil, so as you shall not know which was the Salt, which was the Oil.

Set your glass in Balneo, and there let it stand in an equal degree of fire, till all your matter be turned white and stick to the sides of the glass, and shine like fishes eyes, and then it is white Sulphur of Nature; but if you proceed to the red work, then divide your white Sulphur into equal parts, reserving one part for the white work, and go on with the other part, and in a new glass well sealed up, set it in Ashes till it be turned into a red colour.

When your Sulphur is thus converted, imbibe it again with equal weight of its soul, dissolving and congealing till it remain in an oil, and it will congeal no more, but remain fixed and flowing.

This then is to be fermented with the fourth part of the Oil of Gold, as is often mentioned before.

We have set down already before of the augmentation in quantity and quality, therefore it is not necessary to repeat it here.

We will now return to the white Sulphur before reserved, that we may set down the manner of the white work.

When you have your red Oil or Soul, if you desire to make the white Elixir, set part of the said Oil in a glass ln Balneo to digest, then take it out and put it into a body, and in a lent fire distill away the spirit or white MERCURY, which you must try, that you may know whether it arise pure without water or not, as you do when you try the spirit of Wine, for if it burn all up, it is well;

if it do not, rectify it so often, till it be without any wateriness at all; then have you

rectified your spirit, wherewith dissolve your white Sulphur, till it remain fixed, and flowing, as you did before in the red work, then ferment it and augment it with the fourth part of the Oil of the white Luminary or Luna, as you did the red, and it will be the white Elixir, converting Imperfect bodies into perfect SILVER.

A COROLLARY.

RIPLEY divided the scope of this work into four operations, whereof the first is the dissolution of the body, the second, the extraction of the Menstruum and the separation of the Elements; the third is not necessary in our work, because we cast away the earth after every distillation, instead of which we use our Salt or heart of SATURN; the fourth is, that there be a conjunction of our Salt as is before described.

HEREAFTER FOLLOWETH THE ACCURTATION OF THE WORK OF SATURN.

The way of extracting Quicksilver out of SATURN is found in ISAACUS, of which I know how to make a special accurtation with his water of Paradise, which I gathered partly from the foresaid Author and others; RIPLEY made his accurtation with Quick-silver precipitated with Gold, and the imbibition with Corrosive water, which I like not, because the Elixir so made will be the greatest poison, as himself confesseth, that it were better for a man to eat the eyes of a Basilisk, then taste that Elixir.

But because I desire to set down this accurtation of Lead alone and his Elements, that no strange body may be added to our Elixir, and also that it may be made a Medicine for all uses: I have found out the way of making alone with the MERCURY of SATURN, and his own proper Tincture; for I make a body of one thing which is a spirit, and make that Medicine with its own proper spirit. Read all the Philosophers, and you shall never find a word of this process, nor will any of the Ancients teach thee how to make the MERCURY of SATURN, which that it may be briefly done, this following work will shew at large in our HOLY GUIDE.

CHAPTER II.

The Medicine, Elixir, Fermentation, Imbibition, Precipitation, Quick-Silver, Saturn, Lead, The Toad.

My great Grandfather CHRISTOPHER HEYDON, saith in a certain Manuscript of his, Levi enim Arte norunt Alchimista Mercurium currentens consicere ex plumbo; that is to say, the Alchemists knew how by

73

an easie Art to make current MERCURY out of Lead; but what Art that was, neither he nor any of the ancients have shewed unto us, QUAERITE, QUERITE, saith the first Alchemist (so PARACELSUS was pleased to say in imitation of him) & invenietis, pulsate & operietur vobis, that is to say, seek and you shall find, knock and it shall be opened unto you; which may rather seem to be the words of an envious Master, then the precepts of a Teacher. But having learned this, I learned to seek, that is to say to read; I read, I knocked, that is, I tried many experiments, although they were repugnant to doctrine and Philosophy, therefore although I almost despaired of that Art, yet because nothing is difficult to the industrious, by often knocking, at last I found it apart, by what means I attained to the Art of such a facility, that is to say, of making Quick-silver of Lead; and when the process is read to the operator, it will be rather rejected then believed; but to the end this Art may be revealed as a great secret, I thought it necessary to speak first of the Instruments necessary in this work, before I come to declare the doctrine, which are three in number, that is to say, a Furnace, a Crucible and a pair of Tongs, as appeareth in the HOLY GUIDE.

CHAPTER III.

The Crucible, the Furnace, the Hole in the Top of the Furnace, the Tongues, the Coals.

Let the Furnace be D., the place filled with Coals E., whereunto put fire and when the Coals are well burnt, so that they give a clear flame and fire, take your Crucible A., well annealed that it break not with the sudden heat, and put therein three ounces of filed Lead, having twelve ounces of MERCURY sublimate well ground, and Salt Armoniack six ounces mixed together, which put upon the filings of Lead into the Crucible A., and when the fire is strong and growing hot, take your Tongs C., and presently take up your Crucible, and put it in B., the hole in the top of the Furnace till you hear a great noise and buzzing, then so soon as you can (least the Quick-silver flie away with the spirits) take away the Crucible with the matter therein, and set it in an earthen dish filled with ashes to cool, and when it is cold strike the lower part of the Crucible, so that the matter of the Lead may fall into an earthen dish, and you shall find your Lead converted into Quick-silver.

This Crucible and Furnace is at large characterized in the HOLY GUIDE.

This work is to be reiterated with new spirits till you have a sufficient quantity of Quick-silver, with which proceed as followeth to precipitate this Quick-silver, that from a spirit it may be converted into a fixed body by fixation.

Take of this Quick-silver so much as you please, and put it to precipitate in a round glass well luted, and set in in ashes to the top of the glass; yet let us stay here a while, that your understanding may be the more enlightened.

Therefore understand that the intention of this work is to fix the spirit, which may sooner be done with the spirit of a fixed body, which before was Homogeneal with the body; and which of its own nature desireth to join again with its body.

Therefore nature requireth that she may be helped by Art in this work, to which the Artist consenting, he administreth to adhere unto; which metal is Gold, which is thus prepared, that it be sooner parted by the Quick-silver and stick thereunto.

Take as much pure Gold as you please, and dissolve it in AQUA REGIS mixed with equal part of ACTUM ACERRIMUM, or LAC VIRGINIS, then set it to digest the space of a day, then put your dissolution into an Alembick, and set it ln Balneo, to distill away the water as dry as you can, and do thus three times, and the third time distill it in ashes, that the Salt Armoniac may sublime. Then put distilled Vinegar upon the matter remaining, and after it hath stood three days in Balneo, distill the Vinegar away in ashes, that all the substance of the Salt Armoniack may sublime; and do thus three times, always putting in new Vinegar, until the Oil of the dissolved Gold remain in the bottom; then take of your Quick-silver three times so much as your Gold, and pour it upon the solution of the Gold, that they may mlx together and be united; then put your Quick-silver with the solution in a round Glass stopped only with a piece of Cotton, and with a stick put it down every day as it doth ascend, and keep your Glass in ashes the space of a month, till your Quick-silver be turned into a red precipitate, then again dissolve it in new distilled vinegar, till the whole substance of the quick-silver be dissolved, and the Vinegar be coloured in a golden colour, then distill away the Vinegar in ashes, and again precipitate the quick-silver, which is in the bottom of a Gold colour, into a red and fixed body; and so have you the MERCURY precipitate of SATURN.

It remaineth now that the body be imbibed with its soul, that this being from a spirit reduced into a body, may again imbibe its soul, that it may be dissolved therewith; therefore put it into a Glass, and add thereto equal proportion of its soul or water of Paradise, and shut Your Glass well the space of five days, till the body be dissolved with the soul.

Then dry it in ashes till it penetrates and flows; and when it is dried, try it upon a hot Iron plate, if it be fixed and melt, if not, imbibe it again with half the weight of its water, and do so till you make it fusible and piercing by imbibing and drying it, and when it will melt in the fire, and penetrate, it is the the stone, and fit for fermentation.

We have said enough of the manner of fermentation in the second Book, and therefore it is not necessary to repeat it here; and so after fermentation it will be the Elixir.

Then it is to be augmented and projected, as is before declared; and thus the work of SATURN is accurtated, of which GEORGE RIPLEY saith:

ADROP is the father of the stone, Sericon his brother, LYMPHA his sister, the earth its mother.

But if you desire to know all the secret of SATURN or Lead, I will set you down one process out of PARACELSUS; when you have well prepared the heart of SATURN, saith he, take two or three ounces of that heart and grind it small with double weight of

Salt-peter, and put it in a subliming Glass, with a head well luted to sublime, increasing the fire by little and little as long as anything will ascend or sublime; thus far PARACELSUS: Now if you find this true, RIPLEY will tell you what you shall do with it, in these words.

When by the violence of the fire in the distillation of the Gum of the Sericon, a certain white matter shall ascend sticking to the head of the Limbeck, like Ice, keep this matter which hath the property of Sulphur not burning, and is a fit matter for receiving form, you shall give it form after this manner by rubifying it in ashes, and when it is red Sulphur give it of its soul until it pierce and flow, then ferment it.

Here I have delivered unto you all the ways and manners of SATURN, which are found in any of the Philosophers Books; to the end therefore that the work may be compleated with a demonstration of this word PLUMBUM PHILOSOPHORUM, as appears in the Practical Compendium of RIPLEY, we say that the Philosophers Lead is not taken for Antimony but for Adrop, being converted into the Gum of Sericon.

It remaineth now that we in order treat of the third termination of this Book; therefore, after we have done with SATURN, it is necessary to speak of JUPITER, viz. Tin; but because there are many other

ways of handling SATURN besides those we mentioned I
therefore we refer the Reader thither, seeing he
followeth his footsteps; for he is the offspring of
SATURN and naturally born from him.

CHAPTER IV.

The Third Table of the Elixir of Iron.

It is not necessary to prefix a peculiar Table
to this metal alone, because it is set down before
this book, nevertheless I will here reckon up its
parts and operations as followeth.

1. Calcination.	5. Putrefaction.
2. Solution.	6. Sulphur.
3. Separation.	7. Fermentation.
4. Conjunction.	8. Elixir.

Exaltation or augmentation and projection are
spoken of sufficiently in the former Books.

MARS being most earthy of all the Planets or
bodies, it is not to be doubted but that it may
easily be reduced into a body with little labour;
and therefore most easily converted into Salt, which
is done by Calcination; therefore we will first shew
his conversion into Salt.

Understand therefore, that hence ariseth a twofold consideration, that is to say, that it be calcined one way into its body or Salt, the other way that the body be prepared for solution by calcination.

The practice differeth but a little, for whether you calcine Iron for its Salt or its Menstruum, one only manner of preparation sufficieth.

That is to say, that you take filings of Iron or Steel, as much as you please, and mix therewith equal weight of Sulphur in an earthen body with a Limbeck well luted thereto, then set it in ashes to sublime till all the Sulphur be sublimed from it, then dissolve the filings which remain in the bottom in AQUA REGIA, and it will be converted into Salt, which will be cleansed from the said water, if you put thereon distilled Vinegar and distill it away; do thus three times with new Vinegar, and you shall have a yellowish red Salt in the bottom, which then is a body to be joined to the soul, which keep in warm ashes till you use it.

Now for the practice of Iron or dissolution, take filings of Iron or Steel, so much as you please, and put it in an Iron distilled with Vinegar, and set it in the flaming fire the space of three hours, then take it out and let it cool;

reiterate this work four or five times, then calcine it with Sulphur, as you did before.

When it is thus calcined, set it to dissolve in a corrosive water, by adding equal weight of our ACETUM ACERRIMUM; and let it stand till it has dissolved so much as it can in the cold, then set it in hot ashes, and let it stand there the space of four or five days, pour off the water and dry which is not dissolved, and again calcine it and dissolve it, and when it is dissolved, so as the water be coloured red, pour it out into a body, and keep it till you have dissolved as much calcined Iron as you please.

Then take all your dissolutions, and with an Alembick distill away the water in Balneo, and put distilled Vinegar upon the matter remaining in the bottom, and let it stand upon it in Balneo the space of seven days; then take out your Glass and filter the dissolution, and then again in Balneo distill off the Vinegar, and in the bottom will remain a thick Oil of the Iron or Steel; but if it be not dissolved to your mind, reiterate your solution in RAYMUND'S calcinative water, but it would be better if it were edulcorated with AQUA VITAE, drawing it away again in Balneo, and so you have your Iron dissolved into a liquor.

Therefore proceed to distillation, that there
may be a separation, and distill it in an earthen
Vessel in a strong fire, increasing the fire as much
as you can, and receive the oil, or soul, or red
tincture of MARS separated from the remaining Feces
by the nose of the Limbeck, which oil is the most
permanent tincture for colouring Sulphurs for the
red work, or for exaltation of all Elixirs in
colour, for it makes it tinge and colour higher.

When you have thus prepared the tincture, then
proceed to conjunction, and work with the Salt
before reserved, taking three or four ounces of the
Salt, and equal weight of the soul.

Then seal it up and set it to putrefy in
Balneo, and keep it there till it pass through all
colours and be white, and then it is Sulphur of
Nature.

Then take out your Glass and set it in ashes in
a greater degree of heat till it be red, then
dissolve the red Sulphur with its own soul, and
again dissolve and fix it; dissolving it in Balneo,
and fixing it under the fire, and so it is prepared
for fermentation.

The fermentation is, as hath often been spoken
of before, with the resolved oil of the Sulphur of
Gold in a fourfold proportion to the Medicine, that

by the addition of the ferment, it may be made Elixir transmuting all bodies.

And note that this Elixir of Iron excelleth all other Elixirs, for it rubifieth more, and tingeth higher, and is better for man's body, for it

prevaileth against the spleen, constringeth the belly and cureth wounds, it knitteth broken bones together, and stoppeth the superfluous Flux of the Courses.

CHAPTER V.

The fourth Table of the Physical and Alchymical Tincture out of the red Lion and Glue of the Eagle, drawn out from the Authors Experience.

It is chiefly to be remembered how we first taught you to dissolve Antimony with our ACETUM ACERRIMUM, which may be also well done if you dissolve it in our calcinative water, and after that Antimony is calcined which we spoke of in the end of the second book; it is also to be remembered that in the end of the book I spoke of the Glue of the Eagle in the sixth Table of the first book; these being remembered, it is to be understood that we attribute no other beginning to this accurtation, except that where before we took the blood of the red Lion and the Glue of the Eagle when they were both destroyed; we now join them sound and not hurt together, that

84

they living may mortify and dissolve themselves, which I have fitly called Corporeal Matrimony, or the Union, for in this wedlock they die together, that they may be vivified in the Celestial Matrimony, therefore it is not to be wondered if this Table differ from the other, for this pertaining to the handling of spirits, the other way teacheth the manner of making the Elixir of bodies; therefore we now come to demonstrate the foregoing Table.

Therefore that I may plainly reveal all things unto you, take Antimony well ground, half a pound, and as much Mercury sublimate, likewise ground, and grind them both together upon a marble, till you cannot know them one from another; then set them in a cold place, that the matter dissolving may drop into a Glass set underneath, for when the matters are well mixed together, then say, that they will both shortly be dissolved when the water is perfectly dissolved, it will be of a greenish colour and loathsome smell.

Put this water with the thick part with it into a Glass, and let it stand the space of three days in a fixatory under the fire, and in short time you shall see your dissolvedness of a brownish black colour, and after, that is to say, in the foresaid time it will be red, something higher then red Lead.

85

Dissolve this calcined matter fit RAYMUND'S calcinative water, and when you have dissolved it all into a red liquor or deep yellow, then is your matter brought well into its Chaos.

Put this liquor into a fit body with an Alembic and receiver, and by distillation separate the red oil or the red Mercury from the white body which remaineth in the earth; and if any matter ascend into the head of the Alembic, despise it not, but trie if it be fixed; and if it be not fixed enough, sublime it till it be fixed.

Whereunto join equal weight of its soul, for the Celestial Matrimony, and always leave out the earth in the bottom if you have any sublimate fixed, if not, take the white earth remaining in the bottom, with which proceed as before is said, and join the white body with the soul; when they are thus joined or married, set them to impregnate and revivify in Balneo, till it pass through all colours, and at last be converted into red, which then is the stone.

The manner of Fermentation, Augmentation, both in quantity and quality, and projection, is spoken of before in other works.

And thus Sons, Brethren and Reader, I have delivered and opened (and also have amended many things) all the secrets of the Ancient Philosophers, whose writings were rather published to conceal the Art, then to make it manifest or teach it; although it pleased HERMES TRISMEGISTUS, the first writer of this Art, both to say and protest that he had never revealed, taught, nor prophesied anything of this Art to any, except fearing the day of Judgement or the damnation of his Soul, for shunning the danger thereof, even as he received the gift of Faith from the Author of Faith, so he left it to the faithful; yet when you read his writings, either in his Smaragdine Table, or in his Apocalypse, of his twelve Golden Gates, and shall find nothing plain or manifest, what will you think of such an Author? Believe me all the Ancients have concealed the secret of their preparations in the gross work, although they writ most famously of the Philosophical operation; therefore I have used my endeavor to trie, for out of their writings I found that the Elixir might be made of the Planets or Metals, and also of mean Minerals, which came more near to a metallick nature, then reading more; I found a certain method amongst them all, as it were with one consent or voice on this wise.

First and principally, that bodies should be made incorporeal, that is to say, discorporated, or discompounded, which then is called the Hyle or Chaos.

Secondly, that out of this Chaodical substance, which is one thing, three Elements should be separated and purified.

Thirdly, that the separated and purified elements should be joined, the man and the woman, the body and the soul, heaven and earth, with infinite other names so called, that the ignorant might think they were diverse, which only were nothing else but water and salt, or the body and spirit or soul, that is to say, white MERCURY and red, which they joined together that a new and pure body might be created in putrefaction, that a Microcosmical infant might be created in imitation of the Creation, that is to say, Sulphur of Nature.

Fourthly, That it should be fed with Milk, that is to say, with its own proper Tincture, and after nourished by Fermentation, that it may grow to its perfect strength.

Having learned these, I begun to practice, and in the practice of every body and spirit, I found diverse errors; but reading more and trying more, at last I found the manner and true way of dissolving all bodies, separating and conjoining them; finding the composition of their secret of secrets, that is to say, LAC VIRGINIS, or ACETUM ACERRIMUM, and RAYMUND'S calcining water, wherewith I dissolved all bodies at pleasure, and perfected the gross work, wherefore I purposed, contrary to the custom of the Philosophers, to reveal the whole work, lest I being envious, should be the Author of error like them; therefore I have added their works to my own experiments and inventions, which are plainly and truly writ, that the Artist need to read no books but mine, for herein is almost all things contained, which are found plainly writ by the Philosophers; and also those things which are found true by my own experience.

Now you have all things methodically in this Art without error, with which by the help of God, you may attain to the end.

Alchemy revealeth and openeth unto us four other secrets.

The first is the composition of Pearls, far greater and fairer then natural ones, which cannot be perfectly done without the help of the Elixir.

The second is the manner of making precious Stones of ignoble ones, by the same Art which we taught before in malleable Glass.

The third is the manner of making artificial Carbuncles in imitation of natural ones, which few or none have spoken of.

The fourth is the manner of making Mineral Amber, of which PARACELSUS hath only writ in his book of Vexations of Philosophers, and in the last Edition of his works in the six of his ARCHIDOXES; but because they cannot be made without the help of the Elixir's, therefore they deserve a place amongst the Elixirs; of the fourth, that is to say, of the virtue or rather the vice of making Amber, I shall handle it coldly, I have reserved the explanation of this Aenigma, till the last place.

Some Alchemical Recipes

The following are a number of "recipes" that were translated from German by Mr. Kjell Hellsøe of Stavanger Norway. He found these notes while attending an alchemical class in Salt Lake City Utah. The author is unknown but a possibility exists that the original was written by one Augosto Pancaldi, who had been at the same school in prior years. These notes should be of high interest to the laboratory alchemist.

CONTENTS

I. TARTARUM

The winestone or Tartarum (Potassium Tartrate) was highly valued by the ancients. From it one may obtain:

A tincture (fixed and unfixed)

An oil (fixed and unfixed)

And an alkahest

The alkahest is, however, not obtained by way of the tincture. There are MANY alkahests. The alkahest from winestone is one of the most important ones, but it is not THE alkahest. It works extremely well and will even extract the sulphur from metals.

THE alkahest (i.e., Philosophical Mercury) will completely dissolve gold. This yields the Aurum Potable.

The raw winestone is called Argal or Argo1. The winestone that we shall be working with is the calcined winestone. (Potassium Carbonate)

II. THE TINCTURE FROM WINESTONE

Raw winestone, scraped from the cask, is calcined and then glowed out in order to burn up all the carbon. This is done until it turns a greyish color. This matter is then leached with water (distilled water is best). The water is then evaporated off, and a white powder is obtained, freed from carbon. The powder is then calcined once again and dissolved in water. This is filtered and the water is evaporated off. The powder, or salt of Tartar, should be well dried. This powder is then extracted with a suitable menstruum (such as acetone) to obtain a tincture. From this tincture, one can obtain the oil by the usual methods.

This tincture is of great importance. It is to be used in a fashion analogous to the tincture of antimony. Antimony is a blood purifier. If, however, there are deposits present, one needs a solvent. This is especially true with the tartaric diseases. These (tartaric diseases) are those accompanied by deposits in such bodily parts as the arteries, kidneys, veins and gall. These deposits are quite common and hence, the tincture of winestone is of great value. In addition to combatting the deposits, it is effective against high blood pressure which is caused by the arteries being too narrow. It is also effective against low blood pressure caused by too narrow arteries. It is important not only to keep the blood clean, but to prevent build-up of deposits.

III. THE ALKAHEST FROM WINESTONE

This alkahest is similar to what we have in the vegetable kingdom, alcohol. This we can "sharpen" or actuate, with ammonium chloride.

Similarly, the product from Tartar is half vegetable, half mineral. We take wine in which there is both an acid and a base. viz. acetic acid and winestone. We shall use both. In this wine the potassium went through a change by which it became shut up. The end product is a salt that becomes volatile during distillation and can be distilled over.

Thus, one takes winestone as it comes out of the wine barrel, or better yet, wine that one has made themselves, and evaporates this until one has a fluid thick as honey. One allows all moisture to evaporate, dries it and then calcine it. The matter will turn black and bubbles will arise. The calcination should be continued until the matter is as white as possible. Then acetic acid is poured on the whitened matter. This will cause foaming. From this salt, one may obtain the alkahest.

Solution: One exposes a salt to the air, not directly in the sun, until it becomes fluid (i.e., the matter is hygroscopic). The best way is to put it on an inclined glass plate, or inside a funnel with filter paper, which sits on top of a bottle or receiver. This will allow that which has already dissolved (per deliquium) to run into the bottle. The dissolution in air is to vivify the totally calcined body by means of the air moisture. This moisture encloses (contains) the Spirit. This method is superior to using dew to dissolve the salt. In rain water there is even less spirit. The fluid obtained by dissolving the calcined winestone is then distilled until the salt is dry. Do not discard the distillate as it is more potent than dew since it contains more life (spirit) within it.

Calcined winestone, which chemically has a different formula (K_2CO_3) than untreated winestone ($K_2C_4H_4O_6$) dissolves in the air into a liquid far better than non-calcined. Experimentation has shown that after calcination, a weight loss of about 66% obtains.

In general, acids can be distilled over but not bases. The volatile winestone salt is a base but it IS capable of being distilled. We refer here to a dry distillation, one that has no fluid added. And the term "rectify" means re-distillation, or repeated distillations.

The alchemical Tartarus will dissolve stones and other calcifications in a controlled manner. This process should not be done too quickly lest the arteries become clogged with dissolved matter.

IV. PREPARATION

℞: Dissolve, in the air, one pound of well calcined winestone, per deliquium. Then, the following operations will be repeated ten times: filtration of the fluid, congelation (evaporation of the water), calcination (this process should take 8-10 hours at incandescence but not allowing matter to melt or flux!) Following the calcination, the matter should cool and be allowed to resolve back into a liquid and this entire process done ten times.

Finally, the matter, after the tenth repetition of the various operations, is dissolved in good distilled wine vinegar.

The solution of salt and vinegar is now distilled, under vacuum, and in a water bath. (B.M.) Caution must be exercised so as not to burn the tender flowers. The distillation should be done in a slow process. This will separate the following fractions:

1. The Phlegma. This fraction is over as noted by the arrival of the first sour drops.

2. The next fraction is distilled so slowly that only one drop comes over every 8 seconds. It is finished when the mixture has been thickened to the consistency of syrup or honey.

3. The strongest part now cones over. The process of distillation is continued until smoke is observed and the bulk of the matter ln the retort is dry.

V. PREPARATION OF THE ALKAHEST OF THE WINESTONE (after K. Digby)

Calcine sufficient of your winestone to yield a pound of salt. During calcination, the winestone should not be allowed to flux nor to turn blue. (govern the fire properly). Take the salt now and liquefy it in the air per deliquium then filter it. Coagulate the filtrate via evaporation. Calcine again and heat to incandescence for 5-8 hours without allowing the salt to flux (melt). Then cool it, liquefy it, etc. and repeat the operation ten times. Dissolve the salt now in a 33% solution of distilled wine vinegar. Extract the vinegar by distillation. This distillation is a wary process involving the use of a Balneum Maria (water bath) and vacuum. The Phlegma is to be drawn off in this manner. The distillation is continued until sour drops come over. Then the receiver is changed and the fire augmented until one drop comes over every eight seconds. This is done until the remaining matter attains a syrupy consistency. The receiver is changed again and distillation proceeds until the matter gives off smoke and becomes, for the most part, dry. The distillate is the strongest part and must now be rectified and then added to the weaker part. The phlegm is kept for other purposes.

One rectifies until there is not the least trace remaining in the bottom of the retort. Each time, the retort is to be dried and cleaned. Although the Spiritus Aceti thus obtained is not overly strong, this is alright as it will work well nonetheless. Now one takes 7-8 ounces of the winestone salt and dissolves this amount in the Spirit. This is allowed to stand until black impurities precipitate out. One then filters, coagulates and calcines as before, but with a less strong fire, for about an hour at a temperature barely at incandescence. The matter is to be ground, while still hot with freshly distilled wine vinegar. The impurities are then removed by filtration, then once again it is congealed and calcines as before. This entire operation is repeated until no more impurities show themselves. This should take seven or eight repetitions.

Then place an ounce of this dry substance into a retort and add 100% spirit of wine to make a thin solution, not merely wet. Seal the retort and allow to digest at a temperature of 37°C for twenty four (24) hours. Following this, distill at low heat.

If, instead of the Spirit the Phlegma comes
first due to the Spirit of the Spiritus Vini having
become bound, then one may proceed in exactly the
same way with the remnant part, the remaining
ounces. If not, continue to dissolve in wine
vinegar, then filter, coagulate and calcine until
the spirit stays with the salt; which it will do in
a short time. Thereafter, one proceeds with the rest
of the matter as they did with the test ounce. One
continues to imbibe, distill and distill with
spiritus vini until the SV comes off exactly as
strong as it was when it was poured on. For herein
lies the secret of its sublimation.

Dissolve the impregnated winestone salt with
the Phlegma of the above distilled wine vinegar or
in a very weak SV. Use only as much as is needed to
make the dissolution complete. Shake well. I Then
the very best and finest (subtle) parts of the
winestone salt (weinsteinsalz) will soon be
dissolved and its unprofitable parts left behind,
for the latter is not readily dissolved. Decant this
solution, filter and distill off the Phlegma or the
weak SV. Then the dry spiritus or the dry water,
will fly up, dry, in the form of the most pure
crystals, like icicles. This is the right volatile
winestone salt and the spiritus vini in the form of
a salt. It is the vegetable menstruum that will
dissolve gold leaf in a gentle heat.

The winestone that remains on the bottom during this sublimation must be added to that which the Phlegma of the wine vinegar (or weak SV) did not dissolve. To this, one adds more SV and after calcining with a fire not as strong as previous, and continues the process of fixation. Dissolve this in the air (per deliquium) several times. Then filter and coagulate it as before, and in three (3) repetitions one will obtain more than during the whole previous process, for the winestone has become changed in its nature. Thereafter, imbibe with SV as previously. You will then be able to fix as much SV as you like and sublime as much quantity of pure and clear crystals as you desire.

When the SV becomes fixed on the winestone salt, it will become as sweet as sugar. But when it is separated from it as indicated above, the winestone will keep its most noble nature and yet will be ready for impregnation with much less turbulent movement.

VI. THE VINEGAR OF ANTIMONY

The vinegar of antimony is an alkahest that will relieve inflammation. It contains the element Carbon, which the initial material did NOT. From whence came this carbon?

Also, the regulus of antimony can be melted into a glass, from which one may obtain a tincture. But this tincture is lifeless. The life having flown when the antimony was melted into the regulus. In the same way, one may take the blood from a living person, which will still contain the life force, and the blood from a deceased person, which will NOT contain the life force. The tincture from the regulus of antimony no longer contains the spirit. This, in the form of the vinegar of antimony, is missing, it having been driven off. However, it is not really known if the vinegar of antimony burns in the same way that sulphur does. Now, from antimony, one may extract a mercury that resembles ordinary quicksilver.

VII. PREPARATION OF ANTIMONY

1. One makes glass, pulverizes it, extracts it with ordinary acetic acid and then removes the acetic acid by careful distillation. One then washes with water and evaporates the water off. Now, distilling this once again, something sour comes over. This has been described by Kerckring in his comments to Valentine as: "The Vinegar of Antimony".

2. Melt glass (or: take melted glass), pulverize it and extract a tincture with alcohol. Then, dry the powder and macerate it with dew or

rainwater. Allow to stand for several weeks, the liquid atop the powder will start to turn sour. When this is observed, decant the water and pour on fresh dew or rainwater. Alternatively, one can separate the vinegar from the water and pour this back on the powder. (See Valentine, page 278)

3. One can also macerate the antimony ore directly with water. Rainwater would be best. However, this method will yield some sulphuric acid in the water.

Question: When antimony ore is macerated with water, the vinegar of antimony will be produced and simultaneously a sulphuric acid obtains due to the natural sulphur inherent in the ore. (This is chemical sulphur being described). Is this acid of sulphur of a poisonous nature or is it an alchemical acid, viz: the Mercury of Sulphur?

Answer: When the ore is distilled in its dry state, we will obtain an Alchemical acid of sulphur and NOT a chemical acid.

Note: Some believe the vinegar of antimony is sulphureous acid.

From pyrites (Iron sulphide) one may obtain the sulphur in the form of beautiful light blue crystals. However, as soon as they come in contact with oxygen, they turn brown and are no longer of any great value.

With antimony ore, the same situation prevails. As long as it is not exposed to air, it has quite different (valuable) properties. But as soon as it is exposed to or comes in contact with, moisture, the poisonous chemical acid of sulphur is produced. From this, it would then be possible to obtain the non-toxic alchemical acid of sulphur, but this would be a complex and difficult process. It would be the wet way.

The dry way is much better in this case. However, one must exert care so as not to burn the sulphur. Even then, the separation is a difficult process. Also, from the purified sulphur one may obtain its sulphur (philosophical) and its mercury. These are both non-toxic. The alchemical sulphuric acid will color litmus paper red and is non-corrosive. This means it can be applied, externally, to wounds. Also, in contrast to the chemical sulphuric acid, it does not dissolve quicksilver.

In the chemical element sulphur, there lies a great secret! One will find there an alchemical acid of sulphur which is NOT one of the known acids of sulphur. There are two methods of obtaining the acid of sulphur:

a. The Dry Way (Via Sicca, Trockene Weg) No Menstruum

b. The Wet Way (Via Humida, Nass Weg), With a Menstruum

If one tries to drive out the acid of sulphur from the antimony ore via the wet way, one will only obtain the chemical acid of sulphur. The alchemical and chemical acids can then only be separated with great difficulty.

On the dry way, one may obtain the innate moisture of antimony by distilling without any menstruum. One then has no problem with admixture from acids of sulphur.

This moisture is driven out by the right Regimen Ignis. One takes the ore without calcining it, without adding any moisture it is digested at body temperature for a longer period of time, several weeks. When a vapor is seen ascending, one then distills it off and both Mercuries are thereby obtained. The one from antimony and the other from the sulphur. This will yield an even more excellent medicine! In this process, a vacuum must be used so that not even the slightest bit of moisture can enter.

VIII. THE PHILOSOPHIC GOLD

This is not merely a virtue or spiritual quality, but rather, a substance, a tangible gold. The Philosophic Mercury is required in its preparation. This substance may be prepared by either the wet or the dry way. That is, with or without a menstruum.

On the dry way, without menstruum, the substance is brought to sweat inside a retort. Phlegm will then appear, a watery substance, the "flood". As soon as this has been distilled over, the hitherto dry and bright substance will suddenly turn black. It does not necessarily require weeks and months, in fact, it may prove that only a day is sufficient!

Subsequently, a smoky vapor develops, which condenses again and an almost golden yellow water drips over. The residuum is then pitch black and dry. When ignited, this residuum turns into a yellow or ochre colored mass (depending on the material it consists of). The distilled water is somewhat volatile and has a curious odor. (vide: Nicholas Flamel) This water has a very penetrating smell, even when the vessel is closed. It must be rectified and yields the rectified MERCURIOUS PHILOSOPHORUM, a clear water, Lac Virginis, Aqua Benedicta, etc.

An oil remains behind, the sulphur. After separation, they are joined together again. Thus one obtains a still darker tincture than before; which after a longer period of time, takes on a deep dark color. This is then, the philosophic gold. (vide. A. Cockren)

This Philosophic Gold is **not** the Aurum Potable. It does not contain any gold. However, it DOES have a high degree of medicinal potency. The substance to start out with, will be found in the mineral kingdom. We are, however, not concerned with the four elements; hence the first substance is not water. The first substance has been mentioned so often that one cannot see the forest for the trees!

From the mercury of antimony, one may also obtain the Philosophic Mercury - this is a rather difficult process. From quicksilver also, one may obtain the Philosophic Mercury. This necessitates using living quicksilver which has not yet been rectified. The MERCURIOUS PHILOSOPHORUM can only be obtained from a metal and not a mineral. It is not made from Stibnite. (vide: Valentine, pg. 319)

The Philosophic Mercury is obtained from a very simple substance. The process is also a simple one. It is needed to confect the Firestone.

Sericon is an antimony compound. Probably the sulphide. When extracted with vinegar, it gives the gum. From three pounds of antimony, one obtains two pounds of gum. (another school of thought indicates Sericon could possibly refer to a lead compound)

NOTES

i. The reference to Valentine page 278 can also be found in the English edition of Waite's translation of "Triumphal Chariot of Antimony". This is Kerckring's note on vinegar.

Similarly, Valentine 310 is found in Waite, page 196 which is Kerckring's reference to Phylades and Orestes.

ii. Reference to Cockren is his "Alchemy Rediscovered and Restored"[8].

---

[8] The R.A.M.S. Library of Alchemy Volume 34.

Alchemical Experiments

These were described to Hans Nintzel by David Ham in late 1985.

LIQUOR SILICIS

Take two parts of an alkali such as Potassium Carbonate (KCO_3) and mix it well with one part of clean, red sand. Place the mixture in a fire clay crucible and subject to an intense heat. The two will fuse together at the right, high heat.

When cool, scrape the matter out. It is Liquor Silicis. This hard matter will readily dissolve in water. Glauber suggests using small river pebbles for this rather than sand. He also indicates that gold may be extracted from this.

EXTRACTION OF GOLD FROM ANTIMONY

Take equal parts of crude Antimony (ore) and Tartar. Grind them together to a fine powder which place in a crucible over a strong fire. The two will melt together into a red glass-like mass. If vapors arise, add more tartar.

Keep the matter in the fire for several days and it will congeal and turn different colors, finally becoming dry and brittle in 3-4 days. Then, drive up the fire until the crucible is red hot.

Keep it in the heat about 12 hours then remove and allow to cool.

When cold, dig out the matter and placing it in a crucible, grind it to a fine powder. You will see golden sparkles! Dissolve the matter in hot water. (David says this will yield *Hepar Sulphuris*, and that this is the Universal Salt and Nitre!) The action of the hot water will precipitate some gold dust to the bottom.

EXTRACTION OF GOLD FROM SAND

Take an ounce or so of sand and top it by a few fingers with hydrochloric acid. The HCl will, in a short time, take on a red tinge. Decant this liquid and save. David opines that HCl is the metallic realm's vinegar and that Nitric Acid is the metallic world's alcohol. (analogy to the plant kingdom)

Now pour on the sand, Aqua Regia and obtain ebullition, hissing and such. (stand back!) Distill off the Aqua Regia into a clean receiver. (or, decant the AR into a vessel, first, then distill). As AR distills, it will "lose its grip" on that "hidden in solution" (gets weaker) and gold will precipitate out.

Note: If you take the red tincture obtained by HCl on sand, and pour it on lead filings, the lead will visibly absorb all the red color. It will (the

lead) turn black. Add more tincture and it too will
be absorbed, the lead eventually turning white!!!!

NOTE: One can also pour HCl on broken up ☿
ore, macerate, decant, wash with water and repeat,
until it is well edulcorated and the HCl, decanted
off, is no longer insipid, as at first, but strong.

Pour AR over the sweetened ☿ and stand back.
Distill off and you will find gold flakes left
behind as AR weakens. (Use cold water)

SALTS OF URINE

Privately communicated to Hans Nintzel

9/22/1985 by Arpad Joo-Calgary.

FIXED SALTS

Collect a supply of human urine, preferably
from a younger, healthy person. By placing in a
warm, dark place, allow it to putrefy. NOTE: This
material is vile smelling! Therefore, do ALL
evaporations, distillations, etc. out-of-doors. This
pernicious odor is penetrating, vile and lasting!

Place the putrefied urine in a large vessel and
place outdoors to evaporate. While the sun may be
hot enough to do this, you will probably need a hot
plate. After some time in the heat, the urine will

110

evaporate down to a goo-like or gummy mass. Now drive the heat up, or otherwise calcine this mass until no smoke is given off. Take the feces and leach with distilled water. Then evaporate off the water and allow the salts to crystallize out. Repeat this process until you have a good amount of these valuable salts. These are the 'fixed' salts of urine.

Collect them carefully. Try a bit of the snow-white salt on your tongue. (Oh go ahead!) But a small bit! They will burn as they are quite fiery. Full of fire! These salts are to be distilled over a very high heat. A propane torch may be needed and a vessel that is quite fire-resistant. Be sure to mix the salts with clean sand or brick powder, etc. **before** putting them on the heat. This will prevent the salts from fusing. In any event, the high heat would cause the salts to "dance" in a lively fashion.

An incredible menstruum will be distilled off. A menstruum so sharp, it will radically dissolve gold. (This is a great secret jealously guarded by the ancients!!!) This menstruum can be mixed with Philosophical Mercury (i.e., Spirit of Lead) and thus actuate or sharpen, the Philosophical Mercury to a high degree. Or, you can pour the Philosophical Mercury onto the salt, per se and distill the P.M.

off. If, in this process, the salt turns gray or dark, it can be calcined to white once more.

VOLATILE SALTS

Pour putrefied urine into a large flask, such as an Erlenmeyer, and set it into a distillation train with a double condenser (in tandem) atop the Erlenmeyer. (NOTE: Dr. David Schein and Bill van Doren suggest an "aludel" type train. A sketch of some ideas along these lines is on the next page.)

Boil the urine as hard as you can and snow-white crystals will start to appear in the condenser. (Graham condensers would be the best kind to use). This is the volatile Salt of Urine.

SALTS OF URINE

Glauber calls these salts his "most secret sal ammoniac"! This materia has a FOUL smell. When heated, it will liquify. When cold, it is a white salt. Glauber also refers to it as: mercurial, and has a process where he mixes it with alcohol and sublimes it. Refer to Glauber's works for more details on this marvelous substance. Now, this salt will also sharpen Philosophical Mercury.

A last warning. **DO IT OUTSIDE!!** There are several tracts that deal with urine. See Jugel's Experiments, for example.

SOME MORE 'SOPHISTICATED' APPARATUS TO COLLECT THE VERY VOLATILE SALTS OF URINE

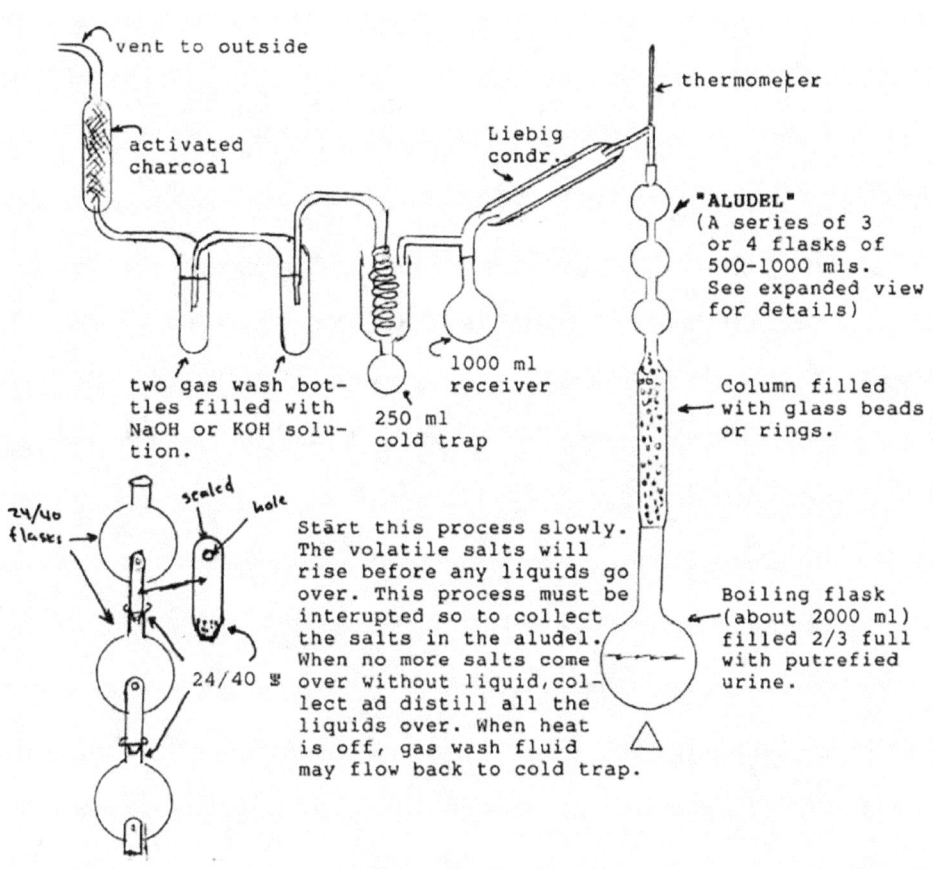

vent to outside

activated charcoal

Liebig condr.

thermometer

"ALUDEL"
(A series of 3 or 4 flasks of 500-1000 mls. See expanded view for details)

two gas wash bot-
tles filled with
NaOH or KOH solu-
tion.

1000 ml
receiver

250 ml
cold trap

Column filled
with glass beads
or rings.

24/40
flasks

sealed

hole

24/40 ₴

Start this process slowly.
The volatile salts will
rise before any liquids go
over. This process must be
interupted so to collect
the salts in the aludel.
When no more salts come
over without liquid, col-
lect ad distill all the
liquids over. When heat
is off, gas wash fluid
may flow back to cold trap.

Boiling flask
(about 2000 ml)
filled 2/3 full
with putrefied
urine.

RED OIL OF ANTIMONY

Communicated Hans Nintzel by Munier Pierre of India, Dec. 1985

Pierre had sent me a photograph of his laboratory and I noted a retort that was colored a bright red. I asked if the color of the glass had some significance. He wrote that the glass was colored due to the action of red oil of antimony that he had made in the retort! I felt that this indicated some sort of great potency and asked for his recipe. This is it.

Materials: pure metal of antimony in powder

cow or bull urine

dried earthworms

distillation devices

Allow the earthworms to dry in the sun so they do not rot or smell vile.

Take equal portions of ♁ metal in powder and dried earthworms in powder. Grind them together into a fine a powder as possible. Now, make a paste of this powder by imbibing it with the urine. (Pierre actually used buffalo urine.) Roll this paste into small balls and allow them to dry in the ☉.

114

When the paste balls are perfectly dry, place them in a distillation train and slowly distill them. A spirit will come over. At first it will be an orange-yellow in color, strong smelling like sulphuric acid. It is very corrosive. This is due to the action of the earthworms and the urine. This spirit is, in itself, a strong medicine and can be made into pills.

Increase the fire now and a blood red oil will come out. A strong fire is needed here, at least, 500°C. The oil will float atop the spirit in the receiver. The oil can be separated and will have a very interesting smell. Strong, but very sweet and fragrant. Keep the oil, once separated, in a flask and it will become thick, like tar. Its color will be darker and it will have a more intense odor. You may well find YOUR retort tinged a deep red by this experiment, or maybe orange-gold.

The distillation process takes 6-9 hours. Ayurvedic teaching indicates that this oil, "projected 21 times on lead" will transmute the lead into 24k gold! The process is to melt some lead, project oil on it, cool it, melt it again, project oil, etc. etc. for twenty one times.

The oil is thick and dark-ruby red, like a gum, with a strong smell. The urine and earthworms clearly open the ♂ radically. Pierre notes there

115

may be oil from the earthworms in the oil of ☿.
(maybe oil of buffalo too?)

DETONATED POWDER OF ANTIMONY

Extracted from "The Triumphal Chariot of Antimony"
of Basilius Valentinus[9] (shown in quotes)

Annotated from experience by Kurt H. von Koenigseck

"The white powder of Antimony is prepared in
the following way: Pulverize equal parts of Antimony
ore (KNO_3) with an equal part of thrice-purified
Saltpetre. (ana[10] by weight)."

The Saltpeter referred to here is crystals of
Potassium Nitrate (KNO_3). I used reagent grade
Potassium Nitrate as produces by Malinckrodt. Any
reagent grade should be pure enough to use.

"When these two matters are well ground
together, burn them in a new-glazed pot, which is
free from all grease, over a circulatory fire, but
not all at once."

This is the process of the ancients called
detonation. Take the two matters and grind them
together very fine. When ready, have a heat-

---

[9] Volume 2, The R.A.M.S. Library of Alchemy.
[10] In equal parts.

resistant dish, a propane torch (or other hot source of fire) in a well ventilated place. It can be done out-of-doors very nicely if no wind.

Get the dish hot and sprinkle some of the mixture in, but no more than half a teaspoon full. The sulphur (S) will begin to smoke and burn off. This should be done, ideally, on a sunny day. Keeping the dish hot, sprinkle more of the powder in. There will be a fizzling type of action. A detonation or very small explosion. Experience will tell how much of the powder to sprinkle on. The dish must ALWAYS be hot. A dark reddish powder should be produced.

To quote Dr. Kerckringius, "It should be prepared under a certain heavenly conjunction, and is the better, the redder it is: for its color is its soul. This is the true **crocus of metals.** Take 8, 9, 10 or 11 grains of this powder, according to the strength of the patient; pour on them 3 or 4 oz. of wine; distill for 4 or 5 hours; the tincture of crocus of metals, which is thus extracted, is like blood. Give it to the patient as a gentle purgative; it will radically cure any disease (!) in the treatment of which it is employed."

To prepare this tincture, separate the white-colored particles from the dark. This is a manual procedure but a pair of pliers can be used as the matter will be rock-hard. Take the white parts and discard them. Take the dark red parts and grind them into a red powder. Wash this red powder with tap water. When the powder sinks to the bottom of the dish, carefully decant off the water and discard. Repeat this procedure until the pH of the wet powder is neutral. (7). Then let the powder dry in the open air. (use distilled water for wash).

When I did this procedure, I noticed a film of oil floating atop the water. This is surely indicative of something!

The "strength of the patient" is purely judgmental. It is referring to size, such as baby, child, teen, adult, etc. A doctor's judgment as to needs would be useful here.

I took 11 grains of powder with 4 oz. of wine and circulated it in a retort for 24 hours. I used "Reunite" Blanco for this. Any good white wine should do. Well, the wine after circulation, was carefully distilled off and poured (cohobated) back on before the powder became too dry. This was repeated 6 or 7 times and then all poured back together and allowed to macerate for a week. The tincture becomes a deep, dark blood-red color. This was carefully decanted. I took 2 mls. Of this tincture at 1 hour intervals until the gentle purgative action started to take place. It served its purpose in a real time of need.

"Now pulverize this hard matter which remains in the pot; pour over it moderately hot water and when the powder has settled at the bottom, add more water. Do this until all the saltpetre has been extracted."

Following each detonation, the powder becomes more subtle and takes longer to settle. This is where care is required if one is not to lose their work down the drain. Also, after each detonation and washing, the powder becomes whiter.

"Dry the substance and again add to it, its own weight (ana) of fresh saltpetre. Burn it (detonate) again. Repeat this operation three times."

During washing, observe the pH, keeping as close to 7 as you can.

"Pulverize what remains, add best spirits of wine and circulate diligently for a month in a well-closed curcurbit or circulatorium; pour to it, and remover spirits of wine 9 or 10 times."

Dry the powder in a heat that is gentle to the touch. The alcohol that I used (in lieu of Spirits of Wine) is "Everclear," which works well. The circulation was done with an up-ended retort, the arm becoming the condenser. When circulation was finished, a receiver was placed on the end of the arm and a slow distillation commenced. Be tilting the arm up, the alcohol ran back, allowing another circulation to begin.

"Gently dry the powder in a crucible such as is used by goldsmiths for melting of silver and gold."

Dry gently until all alcohol is evaporated. (Room temp. will do.)

"Allow the powder to dissolve (per deliquiem) in a humid place or in a hard-boiled egg. (remove yoke from a hard-boiled egg which has been sliced lengthwise. Make sure the pellicule is not disturbed)"

Heat the powder gently so that it does not flux. You do NOT want the powder to fuse, but, rather, obtain a fine-grain structure!

"Dry the liquid which the dissolution (on a marble or in an egg) produces and restore to a pulverized condition."

This powder, if put inside an egg and kept in a humid environment, will turn into a semi-fluid which will easily pour out. Gentle scraping will remove all the wet powder and leave the egg intact for a second resolution. I pour the fluid powder on a piece of plastic wrap from which it is easily removed once it is dry.

Grind the dry powder, again, and place in a covered container. Kirckringius states that: "a dose of a quarter drachm, with generous wine, five or six times, will do the work." My **own** experience has been five (5) grains in the morning and 2 ½ grains in the evening is satisfactory.

Note: I believe the textual wording "five or six times" is a mis-print. I feel he meant to indicate that the dosage is to be continued for five or six **months**, possibly longer, depending on the condition and how fixed it is.

For a Fixed Medicine, circulate the powder with **vinegar** rather than with Spirits of Wine. (6 normal acetic will be fine). Do until all color is extracted out of the powder. Decant off the vinegar from the dregs. Pour all tincted vinegars together and evaporate. Or distill.

When a gum is obtained, wash with warm water. When it is edulcorated, extract with grain alcohol.

SUMMARY

1. GRIND ANTIMONY ORE AND SALTPETRE TOGETHER.

2. SPRINKLE 1/2 TEASPOONS OF IT INTO HOT DISH TO DETONATE.

3. PICK OUT WHITE CHUNKS WITH PLIERS, WASH WITH WARM WATER.

4. GET OUT ALL THE SALTPETRE. GRIND WITH EQUAL WEIGHT SALTPETRE.

5. REPEAT PROCESS THREE TIMES.

6. CIRCULATE WITH SPIRITS OF WINE OR EVERCLEAR[11] FOR A MONTH.

7. COHOBATE SV BACK ONTO P0WDER, RE-CIRCULATE. DO 9-10 TIMES.

8. DISTILL OR EVAPORATE THE SV TO GET A DRY POWDER.

9. LET POWDER RUN PER DELIQUIUM IN A HUMID PLACE OR EGG.

10. DRY THIS LIQUID AND GRIND. THIS IS THE POWDER YOU WANT.

Finis.

---

[11] A brand of Ethanol, 190 proof.

The Oil of Antimony

The techniques given here involve making glass of antimony and have been extracted from Lawrence Principes paper in AMBIX and from his correspondence with Hans Nintzel. The purpose of this paper is to set down the whole process, utilizing the findings he made as he worked with Basil Valentine's instructions, in a convenient single source of instruction to make the oil. A recipe to use! Dr. Principe was kind enough to review and correct this paper.

1. Have on hand Antimony Trioxide. This can be "store bought" or made from Stibium Trisulfide (ore). In either event, one must add some iron oxide in the amount of a half gram to each 100 grams of sulfide (or the oxide if using commercial oxide or pre-made oxide).

2. Take 22 grams of Antimony Trioxide and add to it 0.3 grams of silica. This enables the glass to form. Enough silica **may** be pulled from the walls of the crucible, but the addition of a little silica insures that the glass will obtain. In addition, 0.5 gram of Antimony Trisulfide should be added to the mix (of silica and Sb_2SO_3) before fluxing. In terms of weight, this is 1.5 – 2.0% of silica and 3 – 4% Trisulfide to the Trioxide.

3. This mixture can now be fluxed following Valentine's directions. One can test for 'maturity' as he suggests, using an iron rod (a nail for example) and when the glass IS mature it will adhere to the rod and be perfectly clear. When it is mature, pour the molten glass from the crucible into a brass or copper plate or dish, as thin as possible. (makes it easier to grind later). No Borax need be used as a fluxing agent. Let the glass cool. It will be a yellow glass but other colors can be obtained. (See "Triumphal Chariot of Antimony"). When cool, grind it quite fine, to a powder-like consistency.

4. Beforehand, prepare the extraction medium. Valentine (probably) used wine vinegar and it was probably 'sharpened' with 'salmiac', according to Principe. Salmiac or Sal Ammoniac, is a mixture of ammonium chloride and ammonium carbonate. It is the ammonium carbonate that is important in this operation.

One can use glacial acetic acid reduced with distilled water to 15 - 20% acid. Or, one can use distilled (to remove color!) wine vinegar. In either case add 1 - 2%, by weight, of ammonium carbonate and stir. The carbonate will convert to an acetate.

5. Put the ground glass of ☿ into a flat bottomed flask and top it with the 'sharpened' extraction media by about 4-5 fingers. Set on a gentle heat such as a Balneum Marie and the extraction will take place, the vinegar coloring. When the color is quite pronounced, decant and reserve the tincture. Add fresh vinegar (the extraction media) and set back on the heat. Repeat this process, collecting the tincted vinegars, until no more color can be extracted from the powdered glass of antimony. Combine all the tincted vinegars and filter them.

6. Put the liquid into a distillation train and distill off the vinegarr gently. When the vinegar is over, left behind is a reddish-yellow powder. Sweeten this powder with distilled rainwater and extract it with ethanol (ethyl alcohol). The feces are of no value and may be discarded.

7. Once the powder has been sweetened and extracted with alcohol (which tincts a red color), the oil can be obtained. This is a matter of gently distilling off the alcohol, in a BM (a water bath) to prevent oil from scorching. The oil thus obtained is a marvelous medicine.

8. Note that *salmiac* (a mixture of ammonium chloride and ammonium carbonate) is known as "Salt of Armenia". Armenia was (is) a volcanic region and this salt was obtained from this area near the Black Sea. **NOTE:** This product may well be what Valentine calls "The Bitch of Armenia". In any event, Dr. Principe's research will enable the experimenter to avoid errors and produce a good glass to obtain the sweet oil of ♂.

9. It might be noted that in August of 1990, newspaper articles cite Dr. Principe's studies of the writings of Sir Robert Boyle. He, Dr. Principe, has discovered a coding technique used by Boyle. The decipherment of which is the Ariadne's cord leading the way through a labyrinth of obscure chemical terms.

The Teaching Concerning Atomic Transmutation

By

Volpierre

1892-1952

Translated into English from a Private Manuscript

By Frater Albertus.

INTRODUCTION

As recently as 1971 while I was teaching Alchemy in Walzenhausen, a beautiful spa in the Canton Appenzell in Switzerland, one of the students approached and asked: "Have you heard of an Alchemist by the name of Volpierre?" My answer was in the negative. "His real name was Nikolaus Burtschell" he continued. "He was born in 1892 and died in 1952 in Mainz on the river Rhine." Again, I had to reply that at that moment the name did not ring a bell. He said, "If you are interested in his works I shall be glad to give you all the information I have about him as I have corresponded with him." During overseas classes my time is always fully occupied. Even after the late afternoon periods students often come to me to discuss some question or other. At that time I could only answer that I would indeed be delighted to have any information he might have about this present day Alchemist.

The student was delighted with my reply and immediately handed me some papers with the injunction to please handle them carefully as they were very valuable to him. Among them was a photograph of a man showing only the upper part of his body; the head was in repose and very prominently displayed. It was a picture of Volpierre on his deathbed. Being pressed for time, I took all the papers he had given me; promised to read them and tell him my reaction. After reaching my hotel room a cursory glance through the papers gave me the distinct impression that here was the work of someone who had really tried alchemical experiments in a different manner. I promised myself a more thorough perusal of this work as soon as time was available.

Things do not always work out to one's liking and regretfully I had to return the papers. However, it was agreed that I should receive an exact photostatic copy. After my return to the U.S.A., the papers, including the original photograph mentioned above, arrived. A further investigation of these papers was delayed due to teaching commitments in the Far East. By this time laboratory problems had arisen in Europe and I was asked to arrange a stop there. The student who had given me the Volpierre papers accompanied me to both Zurich and Stuttgart. He inquired about the manuscript of Volpierre; I had to admit that nothing had been done. I determined to do some work with the manuscript on my return home. Further investigation proved it to be most interesting and revealing. Included with the manuscript was a letter with a brief sketch by the student Heinz Fischer-Lichtenthal, of Bavaria, Germany from which I quote:

"In your book *Practical Alchemie in the 20th Century* you mentioned several contemporary individuals who worked in Alchemy. It might interest you that there were others like G. W. Surya (Georgiewic Waizer.) In 1949 I corresponded with another practical Alchemist who belonged to a society of which I also was a member. His name was Nikolaus Burtschell but he used the alchemistical pseudonym D. L. Volpierre. Gathering all that could be found about him, I condensed it into a biographical sketch. A manuscript about his work is in my possession and also the contents of a letter addressed to the editor of a French periodical written in 1948, wherein he strongly defended laboratory alchemy.

In his preliminary sketch, *An Alchemist of the Twentieth Century* he wrote: "The man I am going to talk about was born in the vicinity of Mainz, where he grew up in comfortable circumstances and gained an early education, which, with other endowments was mainly responsible for his ability to read fluently at the age of five. He was a book worm all his life and metaphysical literature was his preference; this was based upon the inheritance from his parents of a treasure of occult experiences. He lived an everyday life but in addition to his daily routine an age old theme came alive within him: Alchemy. When the subject of Alchemy came up and he found it necessary to say a word about it he used the pseudonym, D. L. Volpierre.

"He was engaged in healing most of his life. One of his specialties was Chromo Therapy (healing through color and light.) Under his direction, medications were produced by a pharmaceutical firm with surprising success. In 1932 he healed several obsessed individuals. In the same year, now forty years of age, he began his alchemistical activity; from his eleventh year he gave much thought to a universal tincture and always had the irrevocable conviction that he would reach this goal. Success was reached after a comparatively short time with a Great Tincture, which he produced himself, and he also succeeded with several different transmutations.

"In his treatise, *The Hermetic Art*, Volpierre revealed his attitude toward Alchemy without transgression. This he sent to Salzburg to the metaphysical author, V. G. Surya, for an opinion. Surya gave him unstinting recognition and underlined the correct hermetical laboratory procedure.

"In his Great Work, Volpierre worked primarily with antimony and iron filings. Avoiding Acids produced synthetically he used only those derived from natural sources as did those who preceded him centuries ago. In 1935 it was impossible to obtain proper acids in Germany because of a new law forbidding the use of raw materials containing arsenic, and especially those containing sulphur from Sicily. When using the new industrially produced acids his experiments were unsuccessful. Another goal, to produce a steel-glass, now became impossible.

"His laboratory at St. Inber Saar was destroyed during World War II. From 1943 he lived with his sister in Mainz on the river Rhine helping in her Naturopathic practice.

"In July of 1950 Volpierre suffered a stroke. Death came early in September of 1953. He was buried in the new cemetery in Bischofsheim by Mainz. His former patients still speak highly of him to this very day.

"Reminiscing, his sister described him enthusiastically as having deep blue eyes, light blond hair, and wonderfully revealing facial features; his beautiful soul reflected again and again in his exterior being.

"Volpierre was reluctant to reveal his Alchemistical work and its importance to mankind which had not relinquished its cruelties. Perhaps specific personal experiences convinced this evolved soul personality that he could not cope with the shrewdness and wickedness of the mundane world. A letter written in 1948 in reply to a French magazine article about the transmutation of the elements, testifies to this. One thing that had emerged in the meantime among progressively thinking men was a certainty that Alchemy is the sovereign foundation of all science but is at the same time an art. It remains so at present and will do so in the future. Freed from its many disguises Alchemy will become an irrevocable basis for a continued evolution toward a genuinely modern mankind."

The writer of this introduction was so thoroughly convinced of what Volpierre had accomplished, and of what he had written in his treatise, The Hermetic Art, that he was eager to duplicate his work.

He wanted to know how to procure the raw materials and asked if I would help him to secure them. I agreed. He is still getting things prepared and underway while in the meantime finding further answers concerning various manipulations. He is deeply engaged in laboratory Alchemy. We hope to have a report about his results at a later date when more news is forthcoming from this writer. In the meantime we shall let Volpierre speak for himself.

-Frater Albertus

THE HERMETIC ART

(A Teaching Concerning Atomic Transmutation)

Motto: **What is exalted is simple**

These lines serve as factual information and are intended primarily for those readers who are free from prejudice, have an incompatible understanding, are endowed with a sensitive touch and have retained the fundamentals and simplicity of their comprehension. Writing this I shall purposely omit archaic expressions of the Alchemists whose understanding became only partly clear to my understanding while I was engaged in my self-induced working methods.

At last, this writing shall be a vindication, not to say a complete justification, of the Old Masters, who would have preferred contempt, persecution, even painful death, rather than to reveal their secret though their bones have long since decayed.

This secret, the formula of how to produce the Philosopher's Stone, was undoubtedly known in some Masonic Lodges to the Master of the Chair-as well as to the genuine Alchemists-but it can be safely assumed that at present not one Lodge in Europe has the formula and the know-how to procure the little or Great Tincture. The reader will realize that in

134

my information given herein, I cannot trespass certain boundaries.

Those to whom the expressions of the Old Masters sound strange and unlikely should keep in mind that a gifted poet may speak phantasy to some while to others he proclaims the highest wisdom when presenting his emotions heart to conceive.

The nature of this writing is not to give absolute directions of how to proceed with the work. The result will have to speak for itself.

Knowledge and know-how are ever the new poles of polarity in the unfolding consciousness of man. They emerge out of the innermost atom of an indestructible faith in God and a sensing of the existence of a Divine Being. Being is eternal strength and eternal adding upon. Being is eternally descending and ascending, the eternal change of life and death, but in such a manner that even that which we call death and decay is in the fullest sense of the word a transformation, or rather, a regrouping of matter as a tangible expression in an imponderable, intangible field of force. Manifested force reveals itself in nature as an endless expression in plant, mineral, metal, in a raindrop or a snowflake, or in animal or man. But what a world of difference we find between a gentle zephyr and a roaring holocaust, what a difference between the gentle air in May and the gnashing madness of a hurricane or a blizzard. Such differences show in earthquakes, on land or sea, often connected with volcanic eruptions.

Innumerable are the forces that are in essence only the expression of but one force. For example, note a tree or a flower in growth, blossoming, ripening and withering when looked upon from such a point of view. Even so is man? Externally corruptive but endowed with inner strength of different degrees as a natural creature. So also are the seasons of the year, passing forms within as eternity confined, as is man's unending cycle of birth and death within a micro and macrocosm. We find on the one side sprouting life, decay and transformation, and on the other side continuous change of form and shape reorganizing again manifold manifestations through but one energy which is continuously creative and eternal. The initiate knows of the existence of this energy-the secret Trinity-be it in sunshine or moonlight, be it in the waters of the earth or heavens-be it in the joyously creative womb of Mother Earth, or in the dry, moist cold, or the warm breath of the wind. To know about it means to be able to have such powers that serve us and follow the outlined paths of creation.

The Magnum Opus, the Great or the Royal Work of the Initiates called Alchemists is in itself not difficult and could be taken care of in a few sentences. The work involved is so manifold, so full of surprises that even so-called proven and established physical laws, enough to fill a thick book, would not justify themselves. No wonder that in our time of "rational" thinking and hurried progress aimed toward externalization, instead of directing forces toward interiorization, very few are able to understand the old alchemistical writings, let alone separate the chaff from the wheat. Many expressions used in the work and most especially those that should identify some objects can hardly be understood. Mysterious identifications, such as: Green Lion, Red Lion, the Coat of the Red Lion, Menstruum, Serpent Diana, Phoenix, Lima, Flying Dragon, Virgin's Milk, Echeneu, etc. sound rather fantastic but close examination reveals an intensively grouped correctness of expression that will be proven in practical application.

The alchemical novice is liable to be misdirected when reading intensively, not to say also extremely far-fetched literature, dealing primarily with the procurement of the acid called by the Old Masters respectively vinegar, alcohol, or wine. The novice, his inner hopes raised because he believes he has finally found the way to the Great Work, finds himself in a trap and lost as he discovers in the continuation of his work that he has fallen deeper and deeper into the abyss of bitter despair and hopelessness.

As to when to begin the actual work, all literature handed down to us, as it originated with genuine alchemists, compares as to a hair. Based upon carefully established observations, considering certain sidereal constellations arising out of well-founded calculations, they all fall into place and need not be considered any further herein. Taken practically and factually any alchemistical work may be begun at any hour or any day. However, the end results differ during various constellations, either quantitatively or qualitatively.

Concerning technical aspects, the modern student has an advantage over the old Masters from the very beginning. The old Alchemists had neither gas nor electricity as we have today, but were solely dependent on coal or wood for their fires. These fires were most difficult to maintain and control. Because of their type of heat the alchemists were forced to employ various baths for their work. For instance, I remember the Balneum Mariae and the sand bath required at a certain stage of the work. The first has become superfluous because it is now possible to regulate the gas flame exactly. These various baths have created a fountain of confusion for the alchemistical disciples. Hence they get caught in their own nets, their enthusiasm diminishes and finally their efforts end in discouraged abandonment.

Furthermore, the old Masters used vessels made of clay, such as retorts. When used singly these vessels were called Philosophical Eggs, with a lid glued to the top in a precarious manner which must have been quite a feat in those days. In our time we have vessels made of strong, almost unbreakable glass. Another expression which novices do not know how to interpret is "woman's and children's work." This phrase found repeatedly in alchemistical literature simply refers to the regulation of the fire and of its various degrees of heat as the work requires. One has to think about the clay retorts filled with various substances of different specific weights and boiling points that had to be separated meticulously in the same retort during the work, offering different resistances under pressure that even today with much more sophisticated equipment requires skill even with an easy to regulate gas flame. How much more skillful must have been the regulation of a coal fire!

How much more must our astonishment grow into a marvelous admiration of the Royal Art of the old Masters, some of whom were called Imperator? a title justly deserved by their having attained a rulership over the Opus Magnus.

Each original work of creation is built upon a formative, shaping activity of three vital principles-male, female, and spiritual. That official science acknowledges only two principles-male and female-be they antipodes of complimentary is of no interest to today's Alchemists because while we are engaged in our work, three principals appear absolutely clear in the form of fundamentally different essentials-our Mercury) or Sulphur, and without which no little or Great Work nor a tinging or a transmutation is possible. When only two essentials are considered any scientifically alchemistical conducted experiments will be doomed to failure, without exception, as has been previously proven. It is important and necessary for the student of alchemy-and to be taken for granted by him-that he be mindful of the three reigning principles in the Cosmic, else he will uselessly sacrifice time, money, physical and mental strain.

According to the concept of the old Masters which was looked upon from a purely external point of view, nature was entirely subject to the male and female principle.

Nature receives her eternal strength from the Cosmos which latter cannot exist-and this should be especially emphasized-without the Universal Spirit of the Creator. I shall spare myself to enter into the differentiation about different active force centers within the Macro and the Microcosmos. This was done by the old Masters and during the past five hundred years especially by the one named Paracelsus.

This introduction was necessary even in a limited scope in order to bring us closer to the thought level of the Masters that weaves itself like a red thread through the entire work and proves itself by the final results.

The Opus Magnus has been separated into three parts:

I. Preparation

II. Principal Work

III. Concluding Work

Each section is a masterpiece and has, of course, parallels in the various realms of Nature, be it in regard to heat, moisture, or dryness, which give rise to the sprouting growth in the various

141

stages of the Work, as has already been mentioned by Hermes Trismegistus.

PREPARATION

(The Preparatory Work)

According to the old Masters the earth came forth out of the Chaos, or the Prima Materia. Our first task therefore is to bring any material substance, with which we are working back into its Chaos, if we use the alchemist's language, or to dissolve and change it into its first or chaotic state. The concept of dissolution is here well enough established. It does not deal with annihilation but with a certain degree of dissolvement or rearrangement where the building up of energy takes place and infuses the entire work with life and forms it; however, only after the so-called corporeal has been reduced. The objective or the contents of the entire work can be expressed in one sentence or guide line of Natural Philosophers, out of which has evolved their sovereign rulership over matter: solve et coagula! (separate and unite.) I shall once more draw attention to the great importance of knowing about the male and the female principles of all matter to be considered for the Opus Magnus. A great deal of this knowledge is transmitted to us by way of astrology: a knowledge which Paracelsus especially emphasizes in his writings. A little example shall suffice: According to age old laws copper, sulphur and the number 6 come under the dominion of Venus, just as iron, sodium chloride (common salt) and the number 9 are related to Mars, and lead, saltpeter and number 8 are under Saturn, etc. It is now up to the disciple of the alchemistical Art to cull from such known and established facts the proper proportions necessary to reconstruct the objectives under question. In addition, the degree of heat is of utmost importance in all three parts. Too much heat will bring failure, since the subtle penetrating fumes, which the old Masters called by the meaningful word Spirit, would escape, and for this reason become useless since the active force has escaped.

The individual stages in this process of preparatory work can be seen by a continuous change of color. In my work the object I used was originally of a dark gray, slightly gleaming color, but which became pitch black, then a bluish black, and then a light gray. Then the color changed to dark brown, light brown, and then to a very light gray: almost white. This was followed by bluish white shading changing over to a transparent ice gray, then into an immaculate white. Thereafter it became clouded and then was penetrated by a delightful bright green like a breath, with everything slowly becoming more subdued until it

become of an olive color. Still later it changed to
a yellowish green. During this time there showed a
completely dark closed circle a certain distance in
from the mass on the bottom of the top part of the
container about two centimeters wide-a strange
phenomenon as we shall see later. This circle moved
slowly closer and closer in a steady motion and
became darker and denser. It moved with remarkable
uniformity and kept the same distance changing
slowly to a blue-black color. It began to penetrate
slowly at the edge of the mass on the bottom, but
from the wall of the container, and was not any
deeper than it was wide. At the same time the mass
grew taller from three quarter centimeter to twenty
centimeters. The dark band rested about fifteen to
sixteen centimeters distance from the bottom, about
two centimeters into the mass, while the depth of
the circle or ring was about one to two centimeters.
After a while the dark blue ring became a lighter
blue only to become a deeper blue again several days
later, and gave off a beautiful radiance. Blood red
lines began to appear, rising slowly from the
bottom, until one day a juice of ruby or blood red
color could be seen-the menstruum, also called Aqua
Fontana, that welled forth all at once from the
philosophical mine. The state of this work the old
Masters pointedly called the Peacock's Tail (Gauda
Pavonis.) It is hardly possible to find a more

fitting name. This red juice is also called the Coat of the Red Lion, since the real Red Lion is concealed thereby, because in the course of the further degrees of the work he devours various substances, even different elements. Within this Coat of the Red Lion our Mercury has his being and is rightfully called Bearer of Light. Menstruum, like similar alchemistical terms, has a double meaning. At one time, for instance, through the manipulation of the Master's hand it becomes a perfectly natural excrement of the Original Chaos called Prima Materia, calx of metals greasy earth or matrix. On the other hand, it becomes a solvent, though not a perfect one. As the work is continued one will see what relationship this menstruum has. This red juice or the Aqua Fontana of Paracelsus has to be decanted carefully, making sure that none of the residue on the bottom comes with it. The residue is further imbibed with wine or vinegar and oil, also called Aqua Fortis or Aqua Pluvialis which later also appears in the main work-though in an essentially different form-until it is completely dissolved. Herewith the preparatory work is ended. One has to be careful when touching this fat earth as it is highly corrosive and should not be mistaken because of its greasy consistency.

The Art of the first part of this work consists of the total destruction of the substance used, no matter of what natural origin it is, so as to change nothing and to have only the menstruum left. It is actually possible through certain manipulations with certain liquids that have been named alcohol, wine, vinegar, or on occasion oil-to obtain a red juice, which coloration by now is pending between a delicate shiny raspberry or ruby red out of any metallic substance. The peculiar thing about this red juice in connection with its own color is a green sheen that appears on the wall of the glass when the container is shaken ever so slightly. This green sheen is the symbol of a "growing strength" which is of decisive importance in the continuance of the work. Should this juice besides its green lustre, show a brownish coloration, however slight, the product becomes utterly useless for any further work. Whoever had the fortune to finish this masterpiece of the preparatory work can now with confidence go to the main work.

Out of the preceding method of the preparatory work can be seen-especially in expectation of the end result that the old Masters understood as destruction. As is previously mentioned, a change-over or putrefaction of a corpse, used deliberately as a parallel by the old Masters because a decaying corpse leaks its own juice or "spits"-an old alchemistical expression. At times the matter actually gives off a fetid stinking odor. When the old Masters used the expression, "To transmute a body into chaos," and knew what was understood thereby the description of the preparatory work begins to make sense. However, this should not be confused with a classification, "out of the chaotic realm." The old Masters classified Nature into various realms, such as:

1. Vegetable

2. Mineral

3. Metal

4. Animal

5. Chaotic

6. Astral

What is meant by these classifications can be clearly seen in the old literature of the Masters. The old Alchemists held closely to the concept of but one essential element or substance as a biological dynamic unit whose visible fundamental forms manifest as the four elements. These should not be confused with the elements of scientific postulation. This principle element was preferably called Phosphorous which in essence means "carrier of light." This has given rise to many erroneous concepts of specific hermetic expressions that led to wrong conclusions and false judgments. This is still the case today. There is no need to name such instances-instead let us be satisfied with facts. By the expression Phosphorus was meant the inner fire or inner light that shines but does not consume itself. Indeed, after it has become fixed it cannot be melted nor consumed by any fire. Thus it was correctly named down to its core. However, how this inner fire was to be obtained and how to hold on to it was information wisely withheld.

In the following *Tabula Smaragdina* of the classical alchemist Hermes Trismegistus, this phenomenal knowledge about the three parts of the work has been expressed profoundly and efficiently:

"It is true, certain and without falsehood, that whatever is below is like that which is above; and that which is above is like that which is below: to accomplish the one wonderful work. As all things are derived from the One Only Thing, by the will and by the word of the One Only One who created it in His Mind, so all things owe their existence to this Unity by the order of, Nature, and can be improved by Adaptation to that Mind.

"Its Father is the Sun; its Mother is the Moon; the Wind carries it in its womb; and its nurse is the earth. This Thing is the Father of all perfect things in the world. Its power is most perfect when it has again been changed into Earth. Separate the

Earth from the Fire, the subtle from the gross, but carefully and with great judgment and skill.

"It ascends from earth to heaven and descends again, new born, to the earth, taking unto itself thereby the power of the Above and the Below. Thus the splendor of the whole world will be thine, and all darkness shall flee from thee.

"This is the strongest of all powers, the Force of all forces for it overcometh all subtle things and can penetrate all that is solid. For thus was the world created, and rare combinations, and wonders of many kinds are wrought.

"Hence I am called HERMES TRISMEGISTUS, having mastered the three parts of the wisdom of the whole world. What I have to say about the masterpiece of the alchemical art, the Solar Work, is now ended."

Aristotle was of the opinion that there were four elements in Nature: Earth, Water, Fire, and Air and that they penetrate all realms of Nature in an ever changing pattern, when active life forms are engaged in their primary function. In this connection I refer again to Hermes Trismegistus when he said in the above much recited text, "The wind carried it in his belly." One has to think about the moisture content of air that can, according to prevailing circumstances, become fog, rain, snow, or hail in its free region and in its eternal round become water. In such conditions and in similar occasions occurring in Nature, the alchemist has to leave any concepts aside that are held by those that consider chemical elements, ds used by scientists, which are contrary to his alchemistical conception.

In rain water, snow, frost, or for that matter in any moisture in the air, we find this agent in the form of three essentials least suspected or even recognized by all of mankind - waiting perhaps for a revival or rebirth to reveal its magical forces. Who is going to decide when the time has come and mankind is ready? It was simply impossible for the old Masters, who called themselves Philosophers of Nature and learned from Nature, to conceive of a material manifestation without a soul force. On the contrary, in their opinion concentrated power within matter was known to them which science today after thousands of years has belatedly come to recognize. This conviction of the alchemist urges itself upon the tongue to be proclaimed by all those who have had such an inner experience. This active power induces out of decaying foodstuffs high vitamin contents, out of plants its healing virtues as well as poisons, also it brings forth from flowers and blossoms sweet smelling mandrels as it reveals in visible forms its strength in various degrees as a never ending supply of all that is essential, for such is the genuine Art of the Alchemists.

First, it is important to have the proper medium prepared as given above in the preparatory work, a medium which is possible to bind the active force to itself, similar to a magnet that attracts iron. According to an old teaching like will be attracted to its own, or as it says subtle to subtle, but like forces will repel each other. As the work continues it will be seen just what peculiar stuff this "menstruum" or Blood of the Red Lion really is and that this blood, to use a well-known word from our old master Goethe, in a different sense has a right to be called a "special juice."

THE PRINCIPLE WORK

It speaks for the exalted Art of Alchemy that the old Masters were in a position to accomplish their Great Wonk with the most simple and efficient equipment that was available at the time. In the Principle Work a retort, also called the Philosophical Egg, was primarily used by the old Masters. So called Renomier equipment is senseless and useless.

The retort is filled about three quarters full. One must watch carefully the degree of fire on heat. The reason for this has been mentioned previously. During this so called distillation by degree, for which the old Masters used the athanor, three liquids will appear in succession. Besides this there will remain in the retort a glass-like powder of a yellowish white color, but sometimes of a green or light blue shading. It is noticeable that three entirely different fluids will appear. One is yellow, two are water clear of which the last shows an oily consistency, i.e., it flows like an oil, although in the preliminary work only two liquids were used as a foundation for the entire process. In the three distilled liquids of which the second is called phlegm this is of no use and is poured away. Again we have the male and female principles. These two agents have been given various names as yellow

and white vinegar, wine, alcohol, root moisture,
growing strength, arsenic, sulphur, salt etc. The
remaining powder is called dead earth, the magnet
mumia or the corpse to be revived again. Thereafter
it has to be buried in an alchemistical sense or, as
others express it, buried in horse manure. This
latter expression is also used in similar
terminology in the preliminary work. One sees an
array of seeming contradictions of formulas and
expressions. The real task of the principle work is,
first, the exact completion 'der Scheidekunst' or
art of separation, and thereafter, to revive the
dead earth during its peculiar way of burial when
its own seed is implanted and when grown shows
peculiar blossoms and flowers on one stem.
Separation and purification, the mortification with
the putrefaction and at the same time a revivication
of the earth, compares to a yearly cycle of Nature
and is also known as a "turning of the wheel of
Nature."

The nomenclature of Hermetic Art may prove to
be a hard task to solve by newcomers to alchemy,
since hermetic means sealed airtight. It is a fact
that a certain part of the work has to be
accomplished in an airtight container if Success is
to be had. Furthermore, the alchemistical Masters

knew of a hermetic union whereby several different ingredients were united or welded into 'one' unit as it happened in this work. Even the most exact chemical analysis will not reveal the three, or to be exact, the four different substances formerly mentioned.

In its own 'grave', of its own natural substance, the dead earth receives the Aqua Pluvialis, the rainwater, also known as water of heaven, and begins under its influence not only slowly to revive, but also to decay in a certain way. It begins to swell and thickens, runs when fully absorbed and feels gooey and sticky. During the distillation an absolutely clear water, the very Aqua Pluvialis, comes over, which is absolutely tasteless and odorless, and can be drunk without scruples. During this process, should any fumes arise that were not noticeable before, the work will then enter into a new phase. It now becomes necessary to pour the yellow or white vinegar upon the solidified mass, but with the utmost care and dexterity, otherwise severe physical damage may occur since this manipulation is not without danger. When these two wines are brought together they become a radical solvent and mix under a soft or not so soft hissing noise into a blood red juice which in a surprisingly short time dissolves the entire earth. The menstruum has appeared again but this

time in a purer form as previously and, as the work
continues, becomes even purer and stronger. The so-
called phlegm that is always poured away because of
its uselessness for our work, will add up to a
considerable amount, usually one fourth to a third
of the third total amount whose origin and timing
during the separation is another physiological
enigma. Contrary to the separation of the phlegm,
the menstruum does retain its quantum from the
beginning to the end of the principle work!

During the fifth to seventh revolution of the
wheel of Nature an extraordinary phenomenon appears.
During the process of distillation the fluids partly
turn into invisible fumes and only condense in the
neck of the retort. Suddenly from the bottom of the
retort a milky fluid will rise and begin to move
freely at the top of the retort like clouds during a
storm, to finally flow through the neck of the
retort into the receiver. From the moment the liquid
begins to rise and gathers at the top there is just
enough time-with the greatest care-to change the
receptacle with a carefully set aside new receptacle
to hold the new separation.

This peculiar liquor is the Virgin's Milk, the
Universal Menstruum, radical solvent, the Flying
Dragon also known as Phoenix and Luna. In this
radical solvent all matter of terrestrial origin can

be completely dissolved while retaining their entire essentials and characteristic virtues, which no other solvent in the entire world can accomplish. A careful observation will reveal that this Milk is a light liquid, like water, in which many tiny scales of a silver-white lustre are swimming. This Milk can be kept indefinitely since the wings of Mercury have been clipped, as the old Masters used to say. As soon as the tiny scales called Escheneii or little fish have been separated the possibility of keeping them indefinitely is gone. Another astounding and unusual fact is revealed. When these tiny scales are kept in a hermetically sealed container, they will disappear and leave no residue or condensation whatsoever. These tiny scales have a strong salty-sweet taste. Besides when tasted this powder, like the water, leaves a pleasant warmth in the mouth without being in the least corrosive. During the continuation of the principal work one surprise follows another. For instance, in a decisive stage of the work during the distillation, the yellow vinegar changes its original consistency. It becomes a clear water leaving a peculiar residue in the retort. This residue separates into two powders of equal size presenting themselves in their own way. They rest, as it is said, upon a chair, separated from each other by a small margin on the bottom of the glass in a perfect elliptical shape with their points in a north-south direction. The one powder is white, while the other is yellowish-the hermetic Salt and Sulphur, showing themselves in an extremely fine or subtle state as a volatile Salt-Sal Volatile, i.e., *quinta essential* - after we become acquainted with the volatile Mercury.

In the concluding work these must be fixed which means to avoid evaporation since they as Mars and Venus, like their brother in spirit, Mercury, did fly away while hermetically sealed in their container.

Thus the principle work is finished. The remaining earth is of no further use to us, since from it the volatile double mercury has been leached.

CONCLUSION

The Concluding Work

From the foregoing it can easily be seen that the Opus Magnus required an enormous measure of observation and concentration, as well as a deep feeling for the secret workings of Nature and the entire Cosmic. One should consider that in such a process hundreds, thousands, or millions of years of continuous natural phenomena has been condensed into an average of nine months by the Master's Art, which strangely enough resembles the ripening of the human embryo. The hardest work, technically speaking, is the principal part, that of proper regulation of the gas flame, while remembering the great difficulties encountered by the old Masters attending their coal or wood fires.

As difficult and dangerous as the work is, one is tempted to make comparisons with certain initiation rituals as practiced in some of the Temples of old Egypt. It is easy to understand the decision of a Master that under no circumstances would he repeat the work because of the great dangers involved.

The preparatory work is similar to the principle work of the old Masters. One particular part is to *solve* and the other of the following concluding work is to *coagula*. Some of the aforementioned steps merge into each other during the preliminary and the principal work, so that an exact limitation of the various concepts within the work becomes impossible. Only after the work has been started, and the proper sequence has been fortunately concluded, will things become clearer to the alchemistical laboratory worker. Through proper contemplation one realizes the correct procedures as described in the Opus Magnus with the individual procedures explained in symbolic expressions and associated pictures of the old Masters.

The main purpose of the concluding work is the coagulation and fixation or fastening of the volatile Mercury duplication. With this fixation the old Masters understood their cementation, placing before the worker in the laboratory an abundance of riddles and difficulties. Here, it deals with the Secret and Art of the Alchemists, i.e., the innermost atom of a tightly enclosed, indissolveable conjunction of such matter which gives the tincture and *quinta essentia* its own characteristic being. This ending crowns the Opus Magnus and thus, after much trouble and corresponding patience and perseverance, the final goal is reached.

Depending on the tincture the interplay of color changes during the concluding work. First the substances in question have to be united in their "natural weight." After a short period the mass will become completely black and show many blisters. This same thing happens also during the preparatory work. With a light shining on this black mass (one will be astounded over its deep dark red color-as a blood-like glow.)

It is hard to imagine the change in color that takes place and which differs only from that in the preliminary work through a wonderful shining pureness and beauty that one is tempted to describe as a symphony of color. This glowing of color rises to such a state in the completed lapis or stone, it does not matter whether it be the little or the big Arcanum, it literally glitters, an unmistakable sign that it is the genuine, true tincture.

Depending on how much tincture is prepared, the stone can, during its first multiplication, be raised a hundred fold and thereafter to ten and twenty thousand times its former strength. Through further work the tincture can be raised to still further increase the power of multiplication. The limit of the power of multiplication in its concentration depends on the elements used and their specific weight. When this limit has been reached this tincture will penetrate glass and be as hard and dense as the best steel.

It is from here that a historically proven fact is mentioned about a malleable glass that can be forged, which has defied all efforts of modern science to rediscover.

The highest known concentration was reached by Pryce of England with a potency of 1,140,000. With a single gram of his tincture he brought about an atomic transmutation of 40 kg. Unfortunately this great Master ended his life by committing suicide because he was denied membership in a scientific society.

When using hard or quartz glass higher potencies can be obtained than that which Pryce had reached, but handling such high concentrations can have some unpleasant conditions as a result.

Since this tincture is not confined to metals but can be used equally as well upon minerals, plants and animal substances, when the time has come, it can open avenues of practical application that are presently unlimited if one considers that within the weight of grams of this tincture, kilogram weights of matter are stored.

D. L. Volpierre

(Aqa)

1932

ATTITUDE TOWARDS THE ALCHEMISTICAL OPUS

(Its contents and procedure, describing how to obtain the *quinta essentia* by the great wet way)

By Heinz Fischer-Lichtenthal

In the first part of the Great Work (Opus Magnum) consisting of three parts, any metal carefully dissolved in acid in an artful manner will yield a red juice (menstruum, Blood of the Red Lion.) In such manner a substance will be returned to a certain degree to its primordial condition (chaos), becoming a useful medium capable of extracting the essential vitality found in all creation and then being able to retain it, i.e., such *agens* that is superior to inherent atomic power as we understand it presently.

In the second part a 'menstruum universal' (radical solvent, Virgin's Milk, Flying Dragon) can be obtained from the red juice and its residue through gradual and repeated distillation wherein all natural essentials can be dissolved while their inherent characteristics are still retained.

It contains the Flying Mercury in the form of tiny silver colored scales, like salt, which evaporate quickly when separated from the liquid. Furthermore, from part of the distillation of the yellow vinegar a residue is obtained. This residue consists of two powders (volatile salts) which form on the bottom of the glass container, in the shape of elliptical particles lying next to each other in a north-south direction. One of them is white (hermetical Salt) and the other of, a yellowish color (hermetical Sulphur.)

The final result is obtained only in the third part when all three mentioned salts are fixed. This is the 'cementation'.

By joining (Cohobation) of such *quinta essentia* which does not, dissolve them, the final tincture will give to it the desired character and becomes the Great Tincture. The latter is identical with the Philosopher's Stone.

Begin by mixing one part powdered antimony trisulphide and six parts iron filings in a five litre (5,000 ml), wide necked, glass bottle with ground joint and stopper. Evenly cover the bottom by not more than 3/4 cm.

Next, three pourings on the antimony and iron filings must take place consisting of 3ml of hydrochloric acid and 2ml of sulphuric acid every 24 hours.

Immediately a strong reaction caused by fuming gases makes it imperative that the container be closed very tightly. No antimony particles must be found between the ground stopper and the flask opening. It is best to moisten the stopper with sulphuric acid. Fasten the stopper with some string using two cork stoppers, one on each side, to tighten the string.

Each time some acid has been poured over the mixture place the glass container in a sandbath at a temperature of 37° - 38°C. Let the bottle slowly cool before adding more acid. After the bottle has cooled carefully pour in both acids very gradually and stopper at once. The glass stopper may be removed by inserting strong string tightly between the stopper and the neck of the bottle, then with a uniform jerk on both sides of the stopper, lift it out. Fast work is essential; none of the fumes must escape. Close the bottle very tightly each time after more acid is added.

Beginning with the fourth pouring there is a change of proportion when adding more acid. It is now 6 ml. hydrochloric acid and 4 ½ ml. sulphuric acid which is to be repeated every fourth day. This schedule must be adhered to no matter what the contents of the container shows. The temperature must be kept the same as before except for the short period when more acid is added. Should little lumps show or the substance have a tendency to stick too tightly into a solid mass, use a glass stirring rod and carefully break it up. Be careful not to break the bottom of the retort. Should the glass show a crack a new vessel must be used and the work started anew.

The change of color is as follows:

Dark gray: slight sparkle

pitch black to blue-black

dark gray

light gray

dark brown

light brown

very light gray, almost white

a bluish white

gray, like ice

pure white

glowing green

olive-green

yellowish green

Then a dark floating ring, 2 cm. thick, ascending, getting heavier and darker, (blue-black) enters the mass, which has risen from the bottom to a height of 15 or 16 cm. The color changes from dark blue to light blue, deep, dark blue, now giving off a radiance. Blood red lines begin to rise from the bottom on the side of the vessel. When after circulating for three or four months the rising red juice has reached the top of the mass, the process has reached its sign of development and is concluded.

This juice or tincture can be carefully poured off in any desired quantity. From now on the acids may be added in the same order as before, but in ten to thirty fold quantities or even more, depending on the available space in the container. The mass will dissolve at once.

This solution is kept in special glass bottles for later use. When not enough containers are available only as much as is needed should be dissolved for the work at hand. It is surprising how much acid is necessary to completely dissolve the contents in the glass container.

When shaking the glass container the juice has to show a green lustre. An additional brown coloration shows that the liquid up to this stage has become useless.

The red juice is a natural separation from the Original Chaos. It contains the main element, the

'carrier of light' (Phosphorus-the inner light-fire
that burns, but does not consume itself) which
manifests in all four elements. Chaos or *prima
materia* is the beginning of matter as we know it.
Strength is increased when the *corpora* is reduced.
By too high a heat the fine penetrating fumes
(spirit) would evaporate.

The preliminaries are the beginning of the work
and these should be commenced if possible during a
fortunate trine such as Mercury trine Jupiter, or
trine Venus, or trine the Sun, or a trine with a
well-aspected Mars.

As much as eighty litres, approximately twenty
gallons of the menstruum, has been obtained out of a
five litre bottle, approximately one gallon and a
quarter.

A 250 ml. retort is filled three quarters full
of Dragon's Blood. Begin the distillation with a
small flame and have in readiness three Erlenmeyer
50 ml. flasks. Each liquid will be collected
separately. Using a low flame, first the yellow
vinegar (Hydrochloric acid) comes over, thereafter
the phlegm which is entirely worthless and can be
thrown away.

When nothing more comes over by the small flame
fine threads, rising from the bottom to the top,

shows on the side of the retort and announces that
the sulphuric acid is about to distill over. The
receptacle is changed again and the heat is raised.
Toward the end the bellows or more oxygen will have
to be used to distill the rest of the sulphuric
acid. The retort can be glowing red during the last
procedure. Then let the retort cool and break it.

Remove the remains from the broken retort,
grind in a mortar and submit for three weeks in a
dish to the moist air, preferably in a cellar away
from any light. It is hygroscopic and becomes slimy
and finally thick like honey. Place this in a 100ml.
retort using a small funnel and glass rod. The neck
of the retort must be washed down with the 'Air-
water' to be certain that the neck is absolutely
clean.

Distillation is started with a low flame. Steam
will form, condensing in the neck, and will flow
out. When the water has distilled out the steam gets
denser, like fumes, and the distillation must be
stopped. Let the retort cool, and then break it.
Remove the contents and grind in a mortar. Place the
contents from the mortar in an evaporating dish, and
pour the yellow vinegar over it. There is no danger
in this. Then carefully pour the sulphuric acid
slowly through a funnel. Take care as it may splash.

The two acids combined turn to red and make a loud noise while doing so. Once again we have the product with which we began. The first part of our work is completed when the red liquid is poured through a funnel into a 250 ml. retort and provision made for a repetition of the same process.

The described procedure must be repeated four to six times until the Flying Dragon appears. This can occur during the fifth, sixth, or seventh distillation. One must be especially careful, because our Phoenix can rise suddenly, boiling to the top and flowing out. Quick and efficient changing of the receiving flask is important for this separation.

After Luna has left us, distillation is continued under the same temperature until next the phlegm-like water distills completely over. Continuing the distillation by a lower heat than formerly used, the sulphuric acid comes over. In the remaining water in the retort a peculiar residue forms on the bottom. Next to each other, but separated from each other, a yellow and white salt will appear in an elliptical north-south formation. Both are volatile and can only be kept in their own water. The same applies to the silver-like scales noticed in the residue.

In case too much moisture has been distilled off the salts and the liquid is not sufficient to cover the bottom of the vessel, some of the distillate must be returned. Without this liquid the volatile Salt and Sulphur would evaporate within twenty-four hours, but in the absence of light it would take longer.

It is recommended during the distillation either to extend the neck of the retort with a glass tube for better cooling, or to use a water cooled condenser.

To use the remaining recovered Sulphur again is not recommended!

Formerly the work was conducted solely with natural ingredients. Acids used by the old Masters with their specific characteristics are now hard to obtain or unknown today.

From a chemical point of view we consider hydrochloric acid, sulphuric acid and nitric acid. The latter we can do without for some of the procedures. Hydrochloric acid and sulphuric acid useable for the work described was produced by the Chemical Industry in Europe as little as a few decades ago in a so-called Lead Chamber Process. Raw materials used were native sodium nitrate from the Alsace in France, native Vitriol (copper sulphate) from the mines in the Harz mountains, or Hungary, etc., and red arsenic-sulphur from Sicily. Common yellow sulphur was found to be useless for producing acid to be used alchemistically.

MUTUS LIBER

(ALTUS)

Wherein all operations of Hermetic Philosophy are described and represented.

Preceded by an Explicative Hypotypose of Magaphon.

Translator's Note

The following is a preliminary translation of Magaphon's French commentary to the Mutus Liber. We are counting on the assistance of skilled proof readers to produce a final improved version where errors of spelling as well as such passages where the idiom remains unclear have been corrected. We then hope that this will present a valuable addition to the presently available selection of material on alchemy in the English language. Magaphon was the pseudonym of Pierre Dujois, one of the greatest French erudites around the beginning of the XXth century. He belonged to the circle around Fulcanelli (Robert Amberlain identifies Fulcanelli as Jules Champagne in his book on Spiritual Alchemy).

The Mutus Liber was first published at La Rochelle in 1677. The author's name was given as Altus, a pseudonym. The Mutus Liber also occurs in Manget's "Bibliotheca Chemica Curiosa" of 1707.

More information may be found in "A Prelude to Chemistry" by John Read, London 1936, page 155 et seq. Recognize that the MUTUS LIBER has no words and is simply a collection of plates. (These should be reviewed first)

Kjell Hellesøe

Stavanger Norway

1985

HYPOTYPOSE

This title, good as it may seem, has not the least pretension. It is all technical, the only suitable and genuine on the subject, because it traces, in its conciseness, the plan of our study. A hypotypose (from ὑπό, under, and τύπος, print, emblem) is an explanation placed under abstract figures. Well, the Mutus Liber is then a collection of enigmatic images.

Around the Mutus Liber an absurd legend has formed itself. One school - which has nothing hermetic except the name - has given this work a reputation of impenetrable obscurity, and as such, worships it as a sacrament, without understanding it. This is an error; even as translating Mutus Liber by the Mute Book, without words, is a philosophical misnomer. All the signs adopted by human industry to manifest thought are words. The Latins - this word suitably intended - call drawing, painting, sculpture and architecture, by means of which the Hierogrammates[12] reserved for the elect the mysteries of science, mutae artes, which means the symbolic arts.

What is then a symbol? Συμβολον is a convention; and Συμβολον a sign of recognition. Hence a symbol

---

[12] Sacred scribes, a lower rank of Egyptian priesthood.

is that which we today call a "code", a tacit system of writing adopted for diplomatic and even commercial correspondence, for telegraphic and semaphoric communications etc.

For an illiterate person, all books are mutus. A volume in Hebrew, Sanskrit, Chinese is a mutus liber, a mute book, for the majority, even though they be instructed in their proper language. One has then, to get used to this very simple idea, that the Mutus Liber is a book like all the others and can be plainly read, once one has the grille.

Moreover, the alchemical works, in verse, prose, Latin, French or any other language are themselves nothing but cryptogrammes. Although written with the ordinary letters of the common alphabet and vocabulary they remain no less indecipherable for those who ignore the key. To tell the truth, among the two stenographic procedures, the one of the Mutus Liber is still the more transparent, for the objective image is certainly more speaking than literary tropes and rhetorical figures, especially in a matter as experimental as chemistry.

While attaching these few pages of commentary to the allegorical plates of the Mutus Liber, we are proposing, without leaving the philosophers mantle, to facilitate the lecture, by a sincere interpretation, to the true inquisitors of science, honest, patient and laborious like the diligent bees, and not to the curious, idle and frivolous, who pass their lives uselessly fluttering from book to book without ever pausing at one to extract the malefic substance.

But what! Grammar, geography, history, mathematics, physics, chemistry and the rest do not become accessible but after long and cumbersome efforts, and who would enter into the "King's Palace" without observing the conventions and submit to the laws of etiquette! A hasty and superficial lecture would not replace an austere and serious study. Even the profane sciences are not to be penetrated and assimilated, but after sustained and prolonged work.

One can object that the University counts some illustrious grammarians, geographers, historians, mathematicians, physicians and chemists, but that one never notices the least alchemist there. And if the Bachelor of Alchemy is unknown, it is because alchemy is a chimera. This argument ad hominem is not without an answer: a thing hidden is not at all to be taken for nonexistent, and alchemy is an occult science; or better: it is occult science in its entirety, the universal Arcanum, the soul of the absolute, the magical resort of religion, and that is why it has been called the Sacerdotal or Sacred Art.

There is in all faiths impressed upon the vulgar by means of an appropriate mythology: Bible, Vedas, Avesta, Kings, etc., a positive substratum, which is the foundation of the sanctuaries of all the cults that are propagated on our globe. This mystery, recognized in the catechism as the appanage of the Pontiffs - who are not the public Dignitaries - is alchemy on all its planes: physical and metaphysical. The exclusive possession of the sacrarium, makes up the force of the Churches; they then also watch over the {Masonic Secret} with a worried and jealous care, aided by a suspicious police and censorship.

We are not advancing anything hazardous, and while these allegations may seem unfounded, because impossible, note that since the invention of printing, the hermetic books have always been published freely with license from the civil and religious authorities. And nothing would really be opposed to the spreading of these libels written in known languages, but for the contents; over such a sign which the greatest school chemists - from Lavoiser to Berthelot - have broken their heads without result. And is not this the place to recall the disdainful address of Artephius and the haughty

warnings of the Adepts who ambiguously declare,
write only for those who know and allure the others!
Thus one makes the "Christ" speak in the Gospels,
and the disciples model themselves after the
"Master". It is nonetheless a real and exact
science, conforming to reason and moreover
rationalistic. At all times there have been
"goldmakers"; the "gentleman glassblowers" who
enjoyed such a high consideration, were
hermeticists. And even in our own days,
transmutation still works miracles. After the
sensational debates a little while ago[13], it was told
- and among what stupor - that the Monetary
Administration should have seized, without any other
form of process, and not without reason, the
production of a contemporary alchemists: "You are
not supposed to know how to make gold" he was told
with a threatening air, and was sent back with free
but empty hands. Is it then prohibited to be a sage,
or is alchemy perhaps a state secret? This should
not lead to the naïve conclusion that the ministers
who succeed each other are acquainted with the
Kabbala. The Kings rule but do not govern, according
to a famous aphorism. And for the moment it seems as
if there still is, behind the screen, some grey
eminence who pulls the strings! The famous "Galetas
of the Temple" is perhaps not as abolished as one

---

[13] This introduction was written before the first World War.

supposes him to be, and he would there have a surprising book to write with watermarks on banknotes and with seals on coins.

But in that case, one says, why has gold become so rare as to nearly paralyze social life? The bars have not evaporated, but have been displaced, and one must keep in mind that they will return to their point of departure by an inverse economic movement. Only, too much slowness in this return can have incalculable consequences.

The politics of nations is regulated by a secret metallic pact, which cannot be violated without entailing the most serious international complications. Paper money is therefore to be issued with much effort, but no longer are gold coins to be minted. And yet, it is not the gold that is lacking: it is openly displayed, and with what splendour, on innumerable shoulders, around wrists, on fingers and even legs whose elegance and beauty sometimes leaves something to be desired. Nothing would therefore be easier for the state than to exchange its paper against the precious matter and to put the "coins" into circulation. This is paradoxical, yet true. There is then behind this momentary eclipse in the value of gold a profound reason based on wisdom. {Gold is whatever is gold worth} says an adage. If coining was allowed to the nations who have exhausted their normal reserves, then over abundance would lead to price-cutting. The paper standard would no longer offer any guarantee and would be broken; this would be the death of business and world ruin. This is why even the "natural" production of gold is limited, even as concessions for new mines are refused and until its extraction at poor yield, river beaches and others.

In the meantime, the hour is near when science will totally reclaim all its rights, and when the occult will again become manifest as it was in

181

former times. The sage **Girtaner** has announced it, while basing his opinion on ignored but certain Laws: "In the 20th century the Chrysopee shall be in the public domain." This considerable event is evidently subordinate to a social status quite different from the one that rules today; but we are strongly going there, the world turns fast; and who can foretell the charter of tomorrow!

However, if alchemy is limited solely to the transmutation of metals, it would doubtlessly be an inappreciable science from the industrial point of view, but equally mediocre from the philosophical point of view. In reality it is not like that. Alchemy is the key to all knowledge, and its complete divulgence is called to overthrow from top to bottom, all human institutions that are based on falsehood, in order to reestablish them in truth.

These preliminary considerations seemed to us to be opportune, before charitably taking the reader by the hand, to lead him through the inextricable meanders of the labyrinth. As our intention is to be useful to the seeker, but since we cannot, in a few pages, write a technical treatise, we must before entering into the material, orient the disciple towards the work which best seems to correspond to the figures of the Mutus Liber. The major part of the manipulations indicated in this collection of symbols, find themselves well enough described by the most notorious philosophers, in "*An Open Entrance to the Closed Palace of the King*", by Eirenaeus Philalethes.

It is not that nothing more could be added thereto. Far from it, on the contrary. The practice of Philaletes, presented under an amiable and persuasive exterior, counts among the most subtle and perfidious fictions of hermetic literature. Yet it does contain the truth, but similar to poison, sometimes conceals its own antidote if one only knows how to isolate it from its pernicious alkaloids. As occasion arises we shall signal the traps, in so far as they present themselves to us along our way.

The Mutus Liber is composed of fifteen emblematic plates, some truthful, others misrepresented, and disposed according to one of those beautiful disorders, that, following the precept of Boileau, is an effect of the art.

The first plate that serves as frontispiece is truly capital. On its comprehension depends the whole success of the Work. One sees there, inside a border formed by two interlacing rosaries, a man asleep on a rock whereon grow some languishing kermes - oaks. A limpid water with metallic reflections flows out from it. Next to the sleeper, on a ladder - the Stairway of the Sages - two angels are blowing their trumpets to awaken him.

Above, a propitious and reposeful night sky: the stars are shining and the moon outlines its horn of plenty.

This initial page should bring with it a criticism not directed towards the learned author, but towards the profane artist, who in reproducing the figures has committed, without suspecting, a grave mistake. And it is already a great achievement merely to notice this, without it being necessary to insist more upon it. The hermetic expositions will warn the disciple, who does not judge it useless to inform himself about it.

The sleeping man is the subject of the Work. What is this subject? Some say that it is a body; others affirm that it is a water. They are both right because a water called {the silver beauty} springs from this body that the Sages call the Fountain of the Lovers of Science. It is the mysterious *selago* of the Druids, the matter which gives the salt (from sel for salt and agere to produce).

The secret of the Magistery is to also disengage its sulphur, and to utilize its mercury, for everything is in everything. Certain artists

pretend to turn elsewhere for this in effect, and we
do not deny that the *hydrargyre* of cinnabar maybe of
some help in the work, if one properly knows how to
prepare it for oneself; but one should only use it
knowingly and at the right time. For us, he who
succeeds to open the rock with the staff of Moses,
and that is not a small secret, has found the first
operatory key. On this steep rock then, will flower
the two roses that hang from the branches of the
sweetbriar, one white and the other red.

One will ask us, and not without reason, what
magic word is capable of extracting from the arms of
Morpheus our Epimenides[14], who really seems to be
deaf to the clamour of the trumpets. This word comes
from God, carried by the angels, the messengers of
fire. It is a divine breath that stirs in an
invisible, but certain, manner, and this is no
exaggeration. Without the concourse of the heavens,
the work of man is useless. One neither grafts
trees, nor does one sow the grain during all
seasons. Everything has its time. The philosophical
Work is called the Celestial Agriculture, and not
without cause; one of the greatest authors has
signed his writings with the name Agricola, and two
other excellent adepts known under the names of
Grand Paysan and Petit Paysan.

---

[14] Religious teacher and wonderworker of Crete, who according
to the legend slept for 57 years (ca. 500 B.C.)

Thus, the disciple will have to meditate a long time on this first plate and compare it with the apologues in vulgar language. May he then be fortunate enough himself to hear the heavenly voice; but let him know ahead of time that he will lend his ear in vain, if he has not nourished himself on the Holy Scriptures.

The second plate is not in the order of operations. It represents the philosophical egg, and yet, up to here there is nothing that could have acquainted us with the elements that must compose it.

In order to give us an idea of this, we must deliberately go over a certain number of symbols.

All eggs comprise a germ, Purkinje's vesicle, which is our salt; the yellow, which is our sulphur, and the albumen, which is our mercury. These are enclosed within a retort, which corresponds to the shell. The three products are here personified by Apollo, Diana and Neptune, the God of pontic waters. Tradition requires that this retort be enclosed within a second one, and this one again within a third one made of wood from an old oak. Flamel expressly says: "Note that oak tree", and Vico, the chaplain of the Lords of Grosparmy and Valois recommends it with no less interest. This insistence is significant, and we must recall that in the first

186

plate, on the rock of the Sages grows the Kermes oak, which is the Hermes of the Adepts, because, in the Hebrew language, K and H are but one and the same letter, taken alternatively one for the other. But here one must be on guard, the mineral kermes leads to the trap set by Philaletes, Artephius, Basil Valentine and many others, and one should not lose view of the fact that the philosophers delight in certain verbal collusions. "Ερμῆς is the artificial mercury that amalgamates the compost".

The size of the egg is of no importance. In Nature, the egg varies from that of a wren to that of an ostrich; but, says Wisdom, in *medio virtus*. Something must also be said about the philosophic glass. The authors speak only little about it, and then with reserve. But we know, by experience, that the best is that from Venice.

It must be of a thorough thickness, limpid, without bubbles. In former times one still used the great glass of Lorraine made by the gentlemen glassblowers; but a good practitioner must learn to make his retort by himself.

The lower figure of this second plate represents an athanor between a man and a woman on their knees as if they were praying, which has led certain feeble spirits to believe that prayer intervenes in the work as a ponderable element. Here it is a non-operating factor. The main thing is to employ expedient materials; but the *elan* of the creature towards the creator can have a favorable influence on the directives, for the light comes from God. One must nevertheless be liberate oneself from these not very effective suggestions. The artist's prayer is yet more his work, often hard, dangerous and incompatible with too white hands. Count then above all on the improbus labor.

The third plate is no longer in its place. It leads us into the empire of Neptune. One sees, frolicking in his waves, the dolphin dear to Apollo and on a boat some fishermen putting out their gear. In another ship a man lies stretched out in a nonchalant position. In the second, a landscape with on one side a ram, on the other a bull, which we shall find back further on and shall study at a more opportune moment. Below on the left a woman holds a basket which is the symbol of the grated lantern of the philosophers; on the right a man throws his (fish) line into the ocean which is found within the third circle (the one that encloses the two others). The third circle is animated by a flock of birds to the left; a siren below and Amphitrite up on top. In the margin the sun and the moon, and hovering over this naustic scene, Jupiter carried by his eagle. This whole figure aims at showing that the operator must deploy all his faculties and put to work all the resources of the art in order to capture the mystical fish, about which d'Espagnet speaks.

The author should first have instructed us how to weave the thread necessary for this miraculous fishing. Let us amend his omission: The weir must be braided in a very fine maze from asbestos, which has the property of being incombustible and of staying unaltered. The device being well disposed in the deep waters, one must furnish oneself with a lantern whose luster will attract the prey into the nets. One may also, following other symbols, employ a line; but the Arcanum is in the preparation of the pocket, and the word is circumstantial, for it concerns nothing less than to catch the golden fish.

One will find the secret of this operation in a classical work under the title Ariadne's Thread, for we cannot review the procedure in a few lines within this restricted framework. Concerning the method of lighting the magic lantern indicated by the basket, it is only described in some very rare works, and in a confused manner. Hence we must say a few words about that.

MUTUS LIBER, IN QUO TAMEN

Certain authors, and not the lesser ones, have pretended that the greatest operatory artifice consists in capturing a solar ray, and to imprison it in a flask closed with the seal of Hermes. This gross image has caused rejection of the operation as something ridiculous and impossible. And yet, it is literally true, to the degree that the image coincides with reality. It is moreover unbelievable that one should not have thought of it. This miracle is accomplished in a way by the photographer when he makes use of a sensitive plate which one prepares in different ways. In the *Typus Mundi*, edited in the 17th century by the Fr. of the Society of Jesus, one sees an apparatus, described also by Tiphaigne de Laroche, by means of which one can steal the Heavenly fire and fix it. One can no longer say that the procedure is scientific, and we candidly declare that we are here revealing, if not a great mystery, at least its application to practical philosophy.

The eagles that are flying to the left, inside the great circle are designating the sublimations of the mercury. One needs from three to seven for the Moon, and from seven to ten from the Sun. They are indicated by the flying of the birds and are indispensable, for they prepare the nuptial robe of Apollo and Diana, without which their mystical union would be impossible. That is why Jupiter, the god who governs the eagle, presides over these operations.

The fourth plate shows how the collection of the *flos coeli* works. Some sheets are stretched out on poles in order to receive the heavenly dew. Below a man and a woman are wringing them to press out the divine liquor, that falls into a large vessel put there to that purpose. To the left one sees the Ram; on the right the Bull.

The flos coeli have put the spirit of the bad puffers to torture. Some have seen in it a kind of magical influx, for them magic is a supernatural power, acquired by concourse with spirits, good or bad. Others, more realistic and closer to the truth, have recognized in the morning dew. The flos coeli is in fact called the water of the two equinoxes, from which one has deduced that it is obtained in the spring and in the fall, and is a mixture of these two fluids. Again others, believing themselves still better informed, would collect this mysterious product from a kind of algae or lichenoid whose vulgar name is nostoc. In the Seven Hints on the Philosophical Work, Eteilla, who was perhaps more worth than his reputation, seems to have obtained some satisfactory results with an analogous moss; but one must read his tract with some good glasses.

194

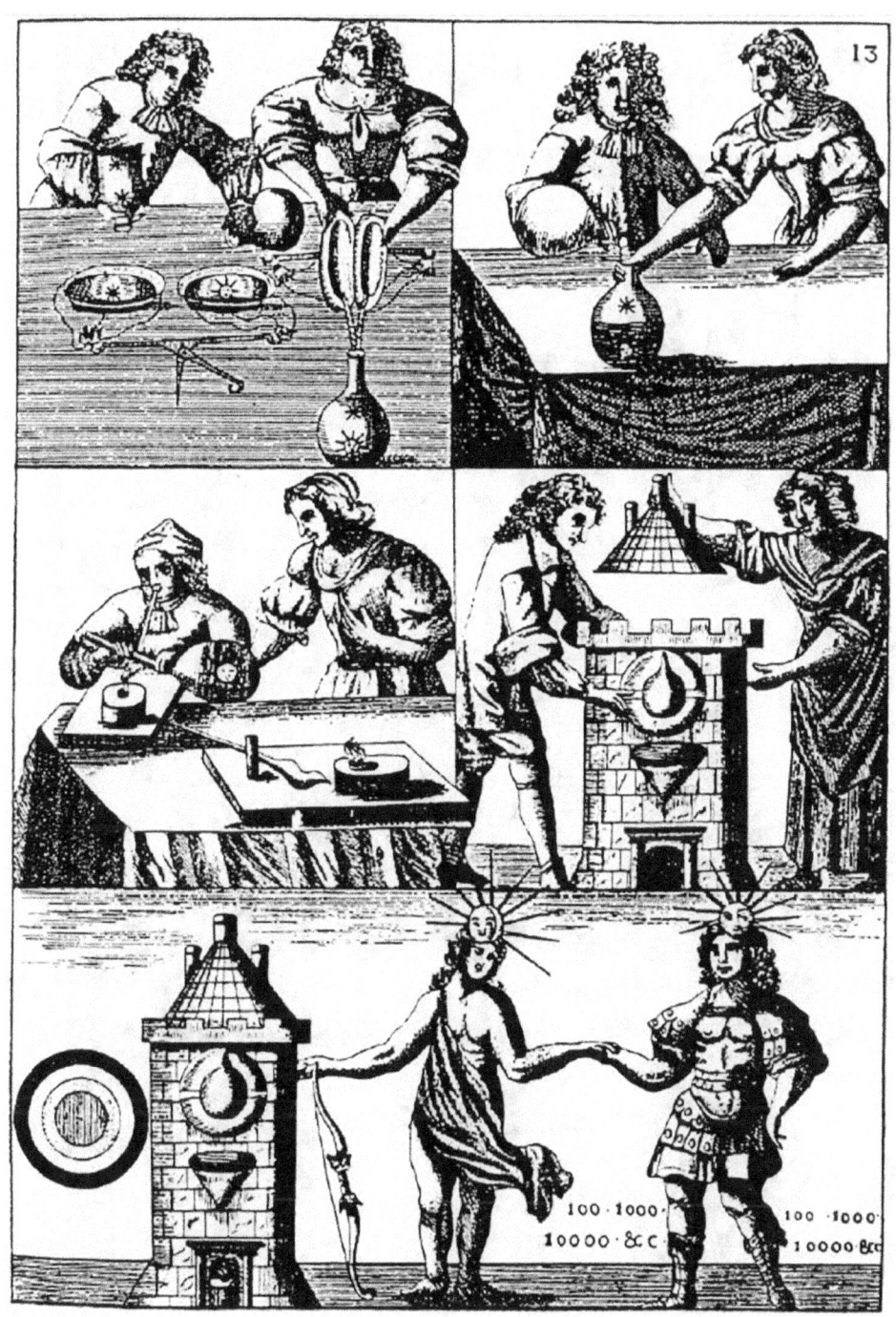

Ora
Lege Lege Lege Relege labora
et Invenies.

196

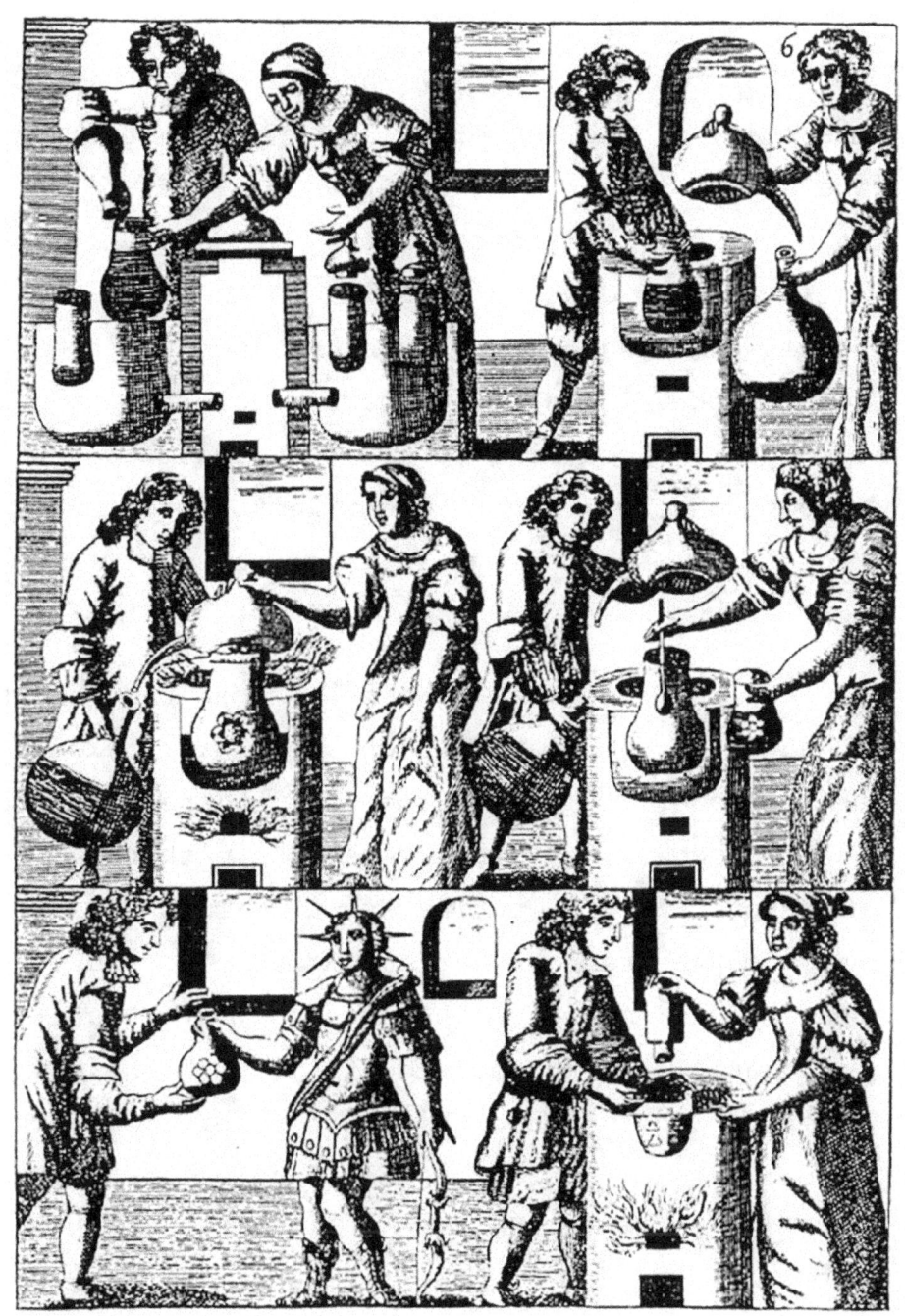

The Rosicrucians called themselves the "Brothers of the ripe Dew" according to the testimony of Thomas Corneille, a good hermetist as was his brother, the great tragic writer. Nevertheless, Philaletes scoffs disdainfully at the collectors of dew and rain water, in which, notwithstanding, the abbot of Vallemont recognized some virtue. It is up to the disciple to form his own opinion according to his own judgement. But it is beyond doubt that an agent kept secret, called "Celestial Manna" plays an important part in the work.

We must declare, sincerely, that the Ram and the Bull in the plate, which one always takes for the signs of the zodiac under which one must collect the flos coeli, have no connection with the astrological symbols. The Ram is Hermes' Criophore, which is the same as Jupiter Ammon; and the Bull, whose horns delineate the crescent, attribute of Diana and Isis, who are identical with the cow Io, lover of Jupiter, and the philosophers moon. These two animals personify the two natures of the Stone. Their union forms the Azyme of the Egyptians, the Asimah of the Bible, hybrid monster designating the *orichalc*, the Latten or bronze oryx, the bull of Phalaris or of bronze, the golden or chrysoeale calf[15],

15 It is not irrevelant to recall here that Helvetius has written a treatise on alchemy under the title Vitulus Aureus.

which differs, most certainly, from the pinchbeck[16]
of Mannheim and is in some ways akin to German
silver[17]. Briefly put, it is the electrum of the
poets; but one should properly understand this word,
which encloses the magical Arcanum. Philaletes
teaches that the gold of the hermetists is certainly
not similar to vulgar gold. Furthermore, we add
that, according to mythology, the Stone devoured by
Saturn was called *betulus*, which is, really, the
same word as vitulus, calf in Latin, and as vitelus,
and the yolk of an egg. The dough of unleavened
bread was its hieroglyph. The priests from the bank
of the Nile never touched the sacrificial breads
with any steel or iron cutting tool: that would have
amounted to a case of sacrilege. From this derives
the ancient custom, still in use, of breaking the
bread. Likewise in the Catholic rite, the
officiating priest divides the host by means of the
silver-gilt paten. This whole logomachy[18] conceals
the vermillion of the Sages, or the philosophic
amalgam of mercury, gold and the silver of art,
rendered indissoluble by the flos coeli.

One will learn, not without surprise, that
bullfighting is a dramatic figuration of the Great
Work. All games have a hermetic origin. The red

(the Golden Calf).
[16] Pinchbeck = copper + zink + tin
[17] German silver = copper, zinc & nickel
[18] Battle of words

cockade worn by the animal and to which it is attached a prize bestowed upon the victor, is the image of the Rose of the Philosophers. The grand concern is to be a good Matador. Also, according to Spanish tradition, to accede to government, one must conquer the bull - the mystic bull, evidently. This victory confers chivalry, true nobility, the one of Science, and consequently the sceptre. This is why, under Louis XIII, the leaders of the "Kabbale d'Etat" were nicknamed Matadors. This species is not extinct, though obliterated and not apparent.

The fifth plate initiates the disciple to the laboratory operations. One there witnesses a sequence of varied manipulations. One will see that it deals with the coction of the fluid collected in the proceeding plate. A man and a woman appear to be pouring it into a pot put over the fire. In the figure below the man adds a viscous product and holds, in his other hand, a substance which is not difficult to discover, if one bears in mind that the egg of Hermogenes is analogous to the others. In the same picture, on the side, a naked person, decorated by a half-moon and embraced by an infant, receives a flask, where one notices four small triangles. They represent the proportions of elements put into the work, namely one part of sulphur to three of mercury. A lunar body intervenes in this operation, which has been indicated by an escutcheon carrying a silver moon on a field of gules.

The moon of the philosophers is not always silver, though this metal is convenient to the work at a certain point. In order to lead astray the profane, the Adepts gave this name to mercury and its salt, the preparation of which presents the most grave difficulties. In order that the mercury be suitable to the work, it is indispensable to animate it. This animation is performed by means of the sulphur prepared to this end. In Philaletes one will find the practical directions, which however must not always be followed word for word. He is exact, yet he fails to purge the mercury from its heterogeneous elements by separating the pure from the impure, the subtle from the gross. One sees, in this plate, the woman who gets ready to skim the compost. It is a presentation overburdened with work, but in the main exact. In the Work it is in fact the feminine element which performs the selection by means of its constitutive virtues; but the artist must lend a hand and assist nature with prudence.

The other figures represent digestions and distillations. We will not be telling the sensible reader anything new, by saying that a man crammed up with chemical formulas and apt to resolve all school

problems on a piece of paper, is not entitled to
call himself a chemist. It is therefore necessary
that theory is accompanied by practice, the one
being the consequence of the other. Only laboratory
practice gives mastership, for what is practice if
not controlled by theory. The rigor of the former
corrects the vagaries of the latter. Thus the
disciple must exert himself to realize all his
concepts.

The sixth plate is the continuation of the
fifth. One will notice that here the operations are
always effectuated by a man and a woman, symbolizing
the two natures. The exterior action of the two
natures indicates the interior work of the mutually
reacting bodies. In the first figure, the female
agent plays a passive role, and the male agent an
active role. The latter is the sulphur, the former:
the moon.

One will doubtlessly want to know which is the
mysterious sulphur concerning which the philosophers
always speak without designating it by any other
name. It is the sulphur of the metals. The secret of
the art consists in extracting it from male bodies
in order to unite it to female bodies, which
requires their previous decomposition. Present day
science seems to consider this fact an absolute
impossibility. But some of the great chemists of the

XVIIIth century have demonstrated, in communications addressed to the academic societies, that the operation is realizable and that they had accomplished it. We have in our hands a magnificent sulphur of silver obtained by analogous means and which closely approaches the tincture of the Sages. But, in order to arrive at this result, a certain practice and profound knowledge of the mineral kingdom is required.

Do not trust the authors who speak of grindings, decantations, separations obtained by what they call a slight of hand. The manual action only contributes to the results in the manner of a kitchen maid preparing her pot-au-feu. When the ingredients have been put into the pot, the water cooks the compost brought to the required temperature by the exterior fire. The coction completed, there remains but to extract the products and employ them according to formula. But all untimely intervention is detrimental and darkness to the work.

Very particularly we must point out the figure which represents the hermetic rose obtained by the foregoing sublimations. On this there would be a lot to say. All alchemical tracts are but (Novels of the Rose)[19], both literally and figuratively. The artists

---

[19] This refers to Jean de Meung's "Roman de la Rose"

foremost concern consists in separating the true from the false. This dominates and constitutes hermetic literature.

What is the Rose? It is the flower of the philosophic tree which forebodes the fruit. Now the tree of the philosophers is the vegetable mercury, the Rose is hence the efflorescence of the metallic sap put into motion by the exterior fire, which excites the internal fire of the bodies. But the Sages speak of two different fires vested with this function. The disciple must therefore consider that there exists, outside of natural fire, another agent of this name, and this secret fire is the ferment of metals, which in the work plays a role analogous to the leaven in the bakers dough. But may the addition of this new element not trouble the minds of the sons of science. For even as leaven is made out of sugar and acidified water, so the ferment of metals is a product of sulphur and mercury, brought to a suitable state by means of art. The proportions are analogous to those used in breadmaking.

Our plate shows a second smaller rose, and a third one smaller still. Would there be several roses? Yes and no. In principle there are two roses, according as one works for gold or for silver; and, basically there is but one. Nevertheless, the Mutus Liber presents three well defined ones. This is correct; but they are daughters of each other, i.e., of three different virtues. During the regimen of coction, Philaletes instructs one to first obtain the white rose, which he calls the moon; then the yellow or saffron; and finally the red or perfect rose. We are not using this author's exact terminology; but we are speaking with enough clarity to make ourselves well understood.

The obtaining of the roses is subordinate to putrefaction. The putrefaction gives rise to a succession of colours. The first is the black; it is the key to the others. Without the black there is no putrefaction; and without putrefaction no transformation. If such an accident should occur, it is because the materials brought into contact do not possess the desired qualities, or have been poorly prepared. See Philaletes for the rest and accept the subtle only.

The seventh plate is very important, but difficult to comprehend. Here we again find the four small triangles with the already explained correspondences; but we are arriving at a delicate operation, for it is here that Saturn devours his child.

The fable of Saturn and Jupiter is well known. But what is this Saturn and what is this Jupiter? The chemical nomenclature, to be found with the authorities, will inform you to what metals these two names correspond. But we remark, in all honesty, that the Saturn and Jupiter of the Sages are not the same as those of the vulgar chemists. One has to be on one's guard, and not try to produce it from the solder of plumber or tinman. We are not working with gross products, and while they have all been derived from the family of metals, they are not proper to the Work before having been submitted to a preparation that renders them "philosophic".

If one adopts the humid way, one proceeds according to art if one brings our two elements in to contact in such a way that one absorbs the other, which gives a new product that will contain the two, without it being henceforth possible to separate them, at least not in a chemical manner. The dry way evidently supposes a combination obtained by a procedure adapted to the nature of the bodies. But one should not mix the two ways: Liquids will unite to liquids and solids to solids.

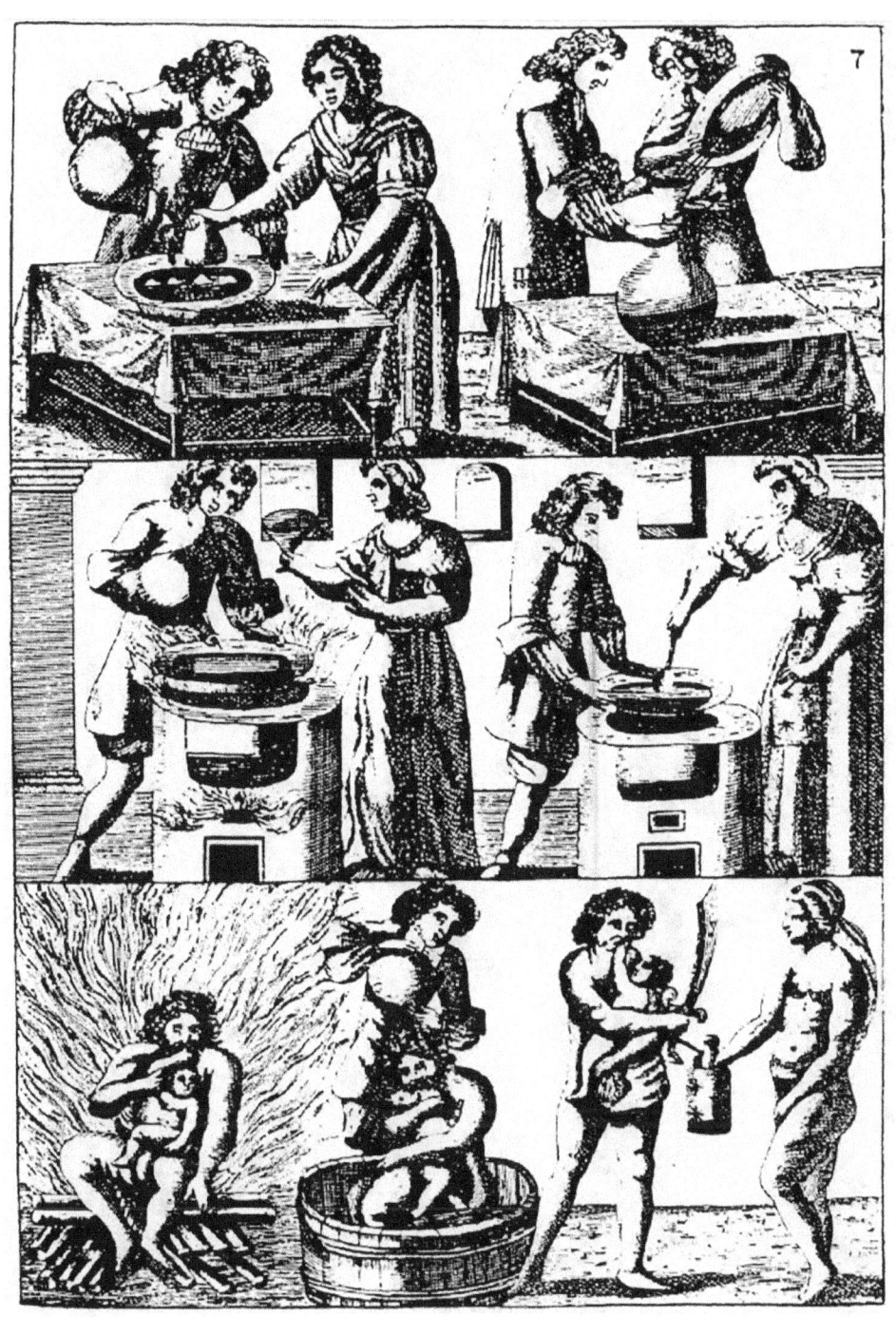

In this operation the fire plays a certain role. One of the figures represents Saturn devouring his son in the midst of a fire. Here one pays the greatest possible attention to the words of the philosophers. One of them will assure you that elementary fire is the destructor of the bodies, and that their fusion volatilizes the soul; another one will declare that the Sages burn with water, and at the same time prohibiting the use of corrosive liquors such as acids. The disciple thus finds himself enclosed within a vicious circle, from which it is extremely difficult for him to escape advantageously. One has to take the mean between the two doctrines in order to make them agree. It is a water which encloses the fire of Heaven; it is the dew, or the flos coeli, which we have seen being pressed out in a previous plate. One knows that the dew contains an acid principle which literally burns. The objects submitted to its action do not delay to turn to dust. We must observe, however, that the philosophic dew in reality differs from the common dew. Nevertheless, it is formed from the true tears of Dawn united to a terrestrial substance, which is the substance of the Work.

When Saturn has accomplished his horrible feast, one must, says Philaletes, cause all the waters of deluge to pass over him, not so that he drowns, but in order to correct the effects of a

laborious digestion and eliminate the toxins
resulting from fermentation. This is what one calls
"to whiten the negro". The operation is rude, but
efficacious, if one preserves, for it has to be
repeated several times. This washing with the noble
water derobes the body of its impurities, corrects
its humors, and disposes it towards the subsequent
operations. It is then distilled hermetically in
order to loose nothing; one precipitates its salt
which presents itself in small and very hygroscopic
crystals, and which must immediately be removed from
the influence of the air. This is why it has to be
shut up, as shown in another figure, in a flask it
has to be shut up, as shown in another figure, in a
flask with a ground stopper, and which one has to
have ready at hand. The eighth plate shows us the
realized philosophic mercury, whereas the second
plate only showed its constituting principles. He is
produced from the Sun and the Moon that are at his
feet. The eagles are flying around him, because
inside the matrass he has to undergo the necessary
sublimations; which have been indicated at the
bottom of the plate by the athanor where the egg has
been placed for incubation. The mercury of the
philosophers, animated and sublimated according to
the rules, must circulate a long time in the vase
before producing the happy effects that one expects
from him. But there are several Mercuries in the

work, and Philaletes points out a second one, very emphatically, under the name of virgin's milk. This one differs in some respects from the first, though they are both of the same essence. Philaletes, Ripley and others so far as to say that it pertains to common mercury. Basil Valentine, on the other hand, banishes it with malediction. Some have believed that the virgins milk could be obtained by combining the two. We are acquainted with one artist who has accomplished this tour de force merely for the pleasure of overcoming the difficulty, without pretending that it has any other advantages. We are thus in a position to acknowledge the operations as possible, which does not imply that we adhere to its use in the actual practice. Only with the utmost reserve should one accept all the bizarre names, imposed by the philosophers upon certain ingredients. These different epithets serve the sole purpose of disguising the course of operations. In this manner the same product carries a different name all according to whether it has or has not been exalted. And it is above all true that alcohol, though extracted from wine, differs from it in name as well as in appearance, in virtue as well as in effect, even as wine differs from the grape from which it has been drawn...

The ninth plate brings us back to the flos coeli. Why this return, and what is the point of repeating it when we were already provided with it? It is not that the author of the Mutus Liber would send us back to the fields to fetch some more; but surely he was obliged to repeat this symbol, the moment that his celestial agent had to enter into a new combination. In one of the figures of this plate, we see Mercury in the process of buying a flask of this divine water from a country woman. He therefore has need for it for some use. Philaletes prescribes, effectively, to wash the mercury several times, in such a way as to make it loose some of its oily nature. He very carefully describes this operation which is accomplished by means of the celestial water brought to a certain temperature, nevertheless moderate, for it takes only a little too much heat for the fiery part to retake its path to the stars. Philaletes is a great master, his word has authority and he presents the work with such ingenuity, that no suspicion of fraud could possibly arise. But we must here expose a ruse: In his work this author has purposely confounded the dry way with the wet way. It would thus be an error to apply to one method that which belongs to the other. But, having made this remark, we recognize that the astral spirit plays a permanent part in the operations. And since we are using an expression of

Cyliani, let us pause at the improbable interpretations to which this very recent term has given rise. Recent writers have seen in this astral spirit a magnetic emanation of the operator. According to them, one must, during a certain period, submit to a physical and moral training, in order to successfully practice this kind of fakerism or yoga. The strength of the product must be proportional to the power of the fluid in such a way that the powder of projection obtained multiplies 100, 1000 or 10,000 times, etc., according to the potential of the artist. Thus these phantasists pretend to impregnate the matter with astral spirit as one charges a battery with electricity. This is what the poorly understood and randomly applied analogy leads to. We shall not name these singular theoreticians whose sincerity is respectable; but we had to signal the fact, in order to put studious and too confident disciple on his guard against the hazardous reading of authors without mandate and without consecration, who have never produced anything but books, but who from that time pass for Masters.

The tenth plate represents the conjunction. The first figure shows, on the scales of a balance, on one side the salt indicated by the star, on the

other the sulphur designated by a flower, which with
its heart forms seven petals. These are the
proportions of the product. A man pours a liquid
enclosed within a flask onto this flower. This is
the Mercury. In his other hand he holds another
receiver filled with astral spirit to be used
according to the circumstances. The woman places all
these products into a long necked matrass; but here
we must recall what we have said concerning the role
of the woman in the Work: the two agents personified
in this way are the matters themselves, and the
diverse accessories that accompany them declare
their state of exaltation.

In the second line, the artist seals the
matrass with the seal of Hermes. He puts its neck
into the flame of a lamp, in order to return the
glass to a pasty and ductible state. Afterwards he
has to draw it out very carefully in order to make
it thinner at the desired point, while assuring
himself that not even the slightest capillarity is
produced through which the spirit of the compost
might escape. Having come thus far, after having cut
the glass, he turns the part adhering to the matrass
back on itself, and shapes it into a thick cushion
(pad). Today this operation is very easily performed
with gas, by means of the blow-lamp. Some very
competent practitioners employ an automatic
procedure of even greater perfection. But whatever

method is used, one thereafter places the egg in the athanor and the coction begins.

We shall say nothing about the athanor. The Mutus Liber presents its form and inner arrangement. Philaletes describes it very carefully. We shall add but one important remark to the sayings of this author: the construction of the furnace is partly allegoric, and there is much to be learned on the point of view of the governing of the fire and of the regimen of the Work. Concerning the latter, the Secret Work on the philosophy of Hermes, attributed to d'Espagnet and favorably quoted, will be useful to follow, for one there finds the Zodiac of the Philosophers.

The last figure of this plate demonstrates that the conjunction is taking place: the Sun and the Moon are united. The work has given the required colors. Here they have been synthesized into a circle, at first black, then white and finally yellow and red. The obtained product multiplies ten times, as announced by the numbers.

The eleventh plate proclaims that the operator has entered the regimen of the Sun, i.e., he has obtained the gold of the philosophers, which is not vulgar gold. Although Jupiter plays a nominal role in the operative process, it is not a question of bisulphur of tin, but of real "mosaic" or secret gold. Meanwhile we shall confess in all truth, that it is not a product of Nature, but of art. Contemporary chemists - unduly taken for competent - have believed to find it in common vitriol, which they hope to render philosophic. They have poorly understood Basil Valentine. The stroma of the dissolution of this salt, considered by them to be "nascent gold", is nothing but a fleeting mirage, and on analysis leaves nothing but deception.

One author, famous with other titles and who in certain circles has enjoyed a certain prestige - we must name Strindberg to warn against his strayings - got stranded on a puerile and ridiculous technique. His Book of Gold is an aberration that calls for a charitable silence. Philaletes and others advise, to those who ignore artificial gold, to seek it in vulgar gold, albeit signaling this work to be long and arduous. One must, in this case, submit it to difficult and dangerous manipulations, for one may transform the metal into gold fulminate, and the memoirs of the XVIIIth century report several mortal accidents following upon this preparation. But if the disciple has been instructed in a good school, he will avoid this sophistic snare and operate hermetically; he will thus avert this redoubtable danger. The masters know how to reach their goal by following other paths, which they take good care not to indicate, but which are not undiscoverable if one reasons with one's reason rather than with the erroneous books of the Sages. "One needs gold in order to make gold", says the classical axiom; this is correct, only there are two different kinds of gold needed to bring the Work to a good end. This plate shows how one here recommences all the preceding operations. One must elevate the mercury to a higher degree of sublimation by means of the eagles, redistill it in order to give it a greater animation.

The twelfth plate instructs us how one may carry this mercury to a superior gradation. To this end one must recommence the imbibitions of the flos coeli, until the mercury, which is eager for them, is impregnated by them to saturation.

The thirteenth plate is a repetition of the tenth, for in the work, all operations follow each other and resemble each other; but this new conjunction, which operates with substances sublimated to the extreme, is nothing but the beginning of the multiplication. The work is the same as in plate ten and, in the coction, one will see the colours reappear. Its duration decreases in proportion as the multiplicative power becomes augmented, in such a way that in the end it takes but one day to obtain the result that in the beginning required months. The numbers on this plate give the transmuting powers obtained by the subsequent coctions.

The fourteenth plate is mainly dedicated to the instrumentation. One there sees the matrass hermetically sealed with its pad, such as we have described it; the mortar and the pestle for the grindings; the spoon for skimming; the balances to determine the right weights; the furnace of the first operations before employing the athanor.

We recall that one must understand the grindings, the decantation, the skimming and all the rest in a philosophic manner, although a trituration, a decantation and skimming are positively necessary to render the substances fit for the work; but, after that, these operations take place by themselves and, so to speak, automatically by the reaction of the bodies on each other. The disciple will have to meditate profoundly on the woman with the distaff, and follow her with sagacity in his manipulations; they are not indifferent and all-telling to the true son of science. We cannot here transgress the will of the author, who witnesses of his well settled design to let the symbol alone express all his thought. If these lines come under the eyes of an adept, he will approve of our reserve, which nevertheless is close to indiscretion. But for the rest: *qui potest capere capiat.*[20]

The fifteenth and last plate represents the apotheosis of Saturn, victorious over his son Jupiter who had dethroned him, and lies inert upon the ground. It is the solarization of the vilest of metals, his resurrection and glorification in the light. The two branches of sweet-briar of the frontispiece are charged with red berries and white berries filled with active seeds where each one has the power to mault all the impure metals into gold or silver. The so-called mystics - who deny the possibility of the metallic work and have found nothing in the allegories of the philosophers but a treatise on ascesis where they would be very embarrassed to explain each symbol - these pseudo-mystics see in this plate an image of the resurrection of man and of his return to the celestial fatherland, and they become blissfully enraptured by this discovery, which they are not far from considering ingenious.

---

[20] He can take it.

But if we become pure spirit again, it follows that our body has enclosed its essence under its (outer) gross form and, given these conditions, one cannot deny the same property exists in the metals. The spirit or the fire is everywhere and in everything: it lies in the flint so cold in appearance, in the metals that one transforms into inflammable fulminates that explode upon the least shock. But, transmutation is a phenomenon that causes the species to pass from an inferior to a superior plane by means of a spiritual agent, a true seed called *powder of projection*. This marvelous product is obtained by the real death and putrefaction of a metallic substance, which, transfigured, has the property of, in its turn, modifying the nature of other matter or creatures. Under its action these likewise undergo a prompt death and resurrection that elevates them to their highest degree of dignity. The Hermetists compare this transformation to the one of wheat. The grain becomes corrupted in the earth, assimilates the gross elements of the soil to itself and, through the work of a long digestion, moults them into pure wheat in the ratio of a hundred to one. This digestion is more or less activated by the environment. In certain climates the harvest takes place three months after sowing, and in the tropics the vegetation obtains almost instantaneously.

It is thus altogether rational that a ferment endowed with great power and projected onto matter submitted to a high temperature may cause it to evolve at a speed that borders on the miraculous.

Evolution is the law of life: the mineral becomes a plant and the plant an animal, by way of intussusception; but this transit is subordinate to the mediation of an exterior agent, plant or beast. If then the metals are admitted to pass in this way from one kingdom to another with the assistance of a suitable element, it is still more logical that a certain perfect and quintessential gold, taken back to its radical and spermatic state has the virtue to exalt and to convert like natures into itself. Is it not thus that the human embryo, during gestation, assumes and transforms the substance of beings of a less noble origin? Nutrition is a continual metamorphosis. Just as, in the three kingdoms, everything converges towards man, among the minerals all ends in gold. But one cannot deduce from this, that nature, in the long run, makes gold from lead. For this effect she needs to be assisted by art, i.e. by the magical ferment that operates its transmutation.

Gold is called the sun, for, in Greek, ἄυρ is light; it is the heaven of the metals, the spiritualization of the species. The metals thus

227

become gold as, in certain regards, our body becomes spirit by the work of posthumous fermentation. Putrefaction, disgusting and hideous, is nevertheless the amazing fairy that works all the miracles in the world. It is a gross error to believe that, in the case of man, the soul abandons the body with the last breath. It is itself entirely flesh, for matter is a modality of spirit in different states subordinated to a greater and more subtle spark, which is the God of each organism. And if science denies the reality of the spirit because it has never found any trace of it, it is dishonouring its own name. A corpse, rigid and cold, is by no means dead in an absolute sense. An intense life, but fortunately unconscious and without perceptible reflexes, continues in the tomb, and it is from this horrible and more or less long combat - which is the purgatory of the religions - that the matter, distilled, sublimated, transmuted and vaporized by the action of the sun, surges up to the amorphous plane, which has its grades from air to elementary light and from this one to the fire principle where all finishes by dissolution and from where all emanates anew.

We believe that we have accomplished our task with all the required honesty, and caused some new light to shine in an obscure domain. It is now up to the disciple to complete the work. As for those who pretend to acquire wisdom without merit and only for a vile and contemptible farthing, we say to them as Saint Jerome in the legend of the rich and idle Cratus: "Philosophy does not fit you".

And you, son of science, bear in mind the eloquent sign by which the terminal figures of the fourteenth plate address you, and the gloss that closes the Mutus Liber: "If you have understood, work in silence and for still some time keep your mouth shut on the Mystery".

MAGAPHON

Quoted Authors and Works

Agricola (Georg Bauer) 1494 - 1595

Artephius

Ariadne's thread (Anonymous)

Marcellin Berthelot 1827 - 1907

Thomas Corneille 1625 - 1709

Cyliani

d'Espagnet - *The Secret Work on the Philosophy of Hermes*

Nicolas Flamel 1330 - 1418

Girtaner

Dr. Helvetius - Vitulus Aureus (1666)

Tiphaigne de Lavoisier 1743 - 1749

Grand Paysan

Petit Paysan

Eirenaeus Philaletes - *An Open Entrance to the Closed Palace of the King*

George Ripley

Typus Mundi - edited by the Jesuits

August Strindberg 1849 - 1912 - *The Book of Gold*

the Abbot of Vallemont

Basil Valentine

the Lords of Grosparmy and Valois

the Chaplain Vico

Boileau

Eteilla - *Seven Hints on the Philosophic Work*

Hermogenes

Finis

Tract on the Tincture and Oil of Antimony

(De Oleo Antimonii Tractatus)

Roger Bacon

On the True and Right Preparation

Of

STIBIUM

To

Heal Human Weaknesses and Illnesses

Therewith and Improve Imperfect Metals

Nurnberg, A.C. 1731

Preface

Dear reader, at the end of his Tract on Vitriol, Roger Bacon mentions that because of the multiplication of the Tincture that is made from Vitriol, the lover of Art should acquaint himself with the Tract De Oleo Stibii. Therefore I considered that it would be good and useful that the Tract De Oleo Stibii follows next. And if one thoroughly ponders and compares these tinctures with one another, then I have no doubt that one will not finish without exceptional profit. Yet, every lover of Art, should mind always to keep one eye on Nature and the other on Art and manual labour. For, when these two do not stand together, then it is a lame work, as when someone thinks he can walk a long path on one leg only, which is easily seen to be impossible,

VALE

Joachim Tanckius

Stibium, as the Philosophers say, is composed from the noble mineral Sulphur, and they have praised it as the black lead of the Wise. The Arabs in their language, have called it *Asinat vel Azinat*, the alchemists retain the name Antimonium. It will however lead to the consideration of high Secrets, if we seek and recognize the nature in which the Sun is exalted, as the Magi found that this mineral was attributed by God to the Constellation Aries, which is the first heavenly sign in which the Sun takes its exaltation or elevation to itself.

Although such things are thrown to the winds by common people, intelligent people ought to know and pay more attention to the fact that exactly at this point the infinitude of secrets may be partly contemplated with great profit and in part also explored. Many, but these are ignorant and unintelligent, are of the opinion that if they only had Stibium, they would get to it by Calcination, others by Sublimation, several by Reverberation and Extraction, and obtain its great Secret, Oil, and *Perfectum Medicinam*.

But I tell you, that here in this place nothing will help, whether Calcination, Sublimation, Reverberation or Extraction, so that subsequently a perfect Extraction of metallic virtue that translates the inferior into the superior, may profitably come to pass or be accomplished. For such shall be impossible for you.

Do not let yourselves be confused by several of the philosophers who have written of such things, i.e., *Geber, Albertus Magnus, Rhasis, Rupecilla, Aristoteles* and many more of that kind. And this you should note. Yes, many say, that when one prepares Stibium to a glass, then the evil volatile sulphur will be gone, and the Oil, which may be prepared from the glass, would be a very fixed oil, and would then truly give an ingress and Medicine of imperfect metals to perfection.

These words and opinions are perhaps good and right, but that it should be thus in fact and prove itself, this will not be. For I say to you truly, without any hidden speech; if you were to lose some of the above mentioned Sulphur by the preparation and the burning, as a small fire may easily damage it, so that you have lost the right penetrating spirit, which should make our whole Antimonii corpus into a perfect red oil, so that it also can ascend over the helm with a sweet smell and very beautiful colors and the whole body of this mineral with all its members, without loss of any weight, except for the foecum, shall be an oil and go over the helm.

And note also this: How would it be possible for the body to go into an oil, or give off its sweet oil, if it is put into the last essence and degree? For glass is in all things the outermost and least essence. For you shall know that all creatures at the end of the world, or on the last and coming judgement of the last day, shall become glass or a lovely amethyst and this according to the families of the twelve Patriarchs, as in the families of jewels which Hermes the Great describes in his book: As we have elaborately reported and taught in our book "de Cabala".

You shall also know that you shall receive the perfect noble red oil, which serves for the translation of metals in vain, if you pour acetum correctum over the Antimonium and extract the redness.

Yes not even by Reverberation, and even if its manifold Beautiful colors show themselves, this will not make any difference and is not the right way. You may indeed obtain and make an oil out of it, but it has no perfect force and virtue for transmutation or translation of the imperfect metals into perfection itself. This you must certainly know.

AND NOW WE PROCEED TO THE MANUAL LABOR

AND THUS THE PRACTIC FOLLOWS

Take in the Name of God and the Holy Trinity, fine and well cleansed Antimonii ore, which looks nice, white, pure and internally full of yellow rivulets or veins. It may also be full of red and blue colors and veins, which will be the best. Pound and grind to a fine powder and dissolve in a water or Aqua Regis, which will be described below, finely so that the water may conquer it.

And note that you should take it out quite soon after the solution so that the water will have no time to damage it, since it quickly dissolves the Antimonii Tinctur. For in its nature our water is like the ostrich, which by its heat digests and consumes all iron; for given time, the water would consume it and burn it to naught, so that it would only remain as an idle yellow earth, and then it would be quite spoilt. Consider by comparison Luna, beautiful clean and pure, dissolved in this our water. And let it remain therein for no more than a single night when the water is still strong and full of Spirit, And I tell you, that your good Luna has then been fundamentally consumed and destroyed and brought to naught in this our water.

And if you want to reduce it to a pure corpus again, then you will not succeed, but it will remain for you as a pale yellow earth, and occasionally it may run together in the shape of a horn or white horseshoe, which may not be brought to a corpus by any art.

Therefore you must remember to take the Antimonium out as soon as possible after the Solution, and precipitate it and wash it after the custom of the alchemists, so that the matter with its perfect oil is not corroded and consumed by the water.

THE WATER IN WHICH WE DISSOLVE OUR ANTIMONIUM

IS MADE THUS;

Take Vitriol one and a half (alii 2.lb.) Sal armoniac one pound, Arinat (alii Alun) one half pound, Sal niter one and a half pound, Sal gemmae (alii Sal commune) one pound, Alumen crudum (alii Entali) one half pound. These are the species that belong to and should be taken for the Water to dissolve the Antimonium. Take these Species and mix them well among each other, and distill from this a water, at first rather slowly. For the Spiritus go with great force, more than in other strong waters. And beware of its spirits, for they are subtle and harmful in their penetration.[21]

---

[21] Wear a face mask! -HWN

When you now have the dissolved Antimony, clean and well sweetened, and its sharp waters washed out, so that you do not notice any sharpness any more, then put into a clean vial and overpour it with a good distilled vinegar. Then put the vial in Fimum Equinum, or Balneum Mariae, to putrefy forty (alii four) days and nights, and it will dissolve and be extracted red as blood. Then take it out and examine how much remains to be dissolved, and decant the clear and pure very cautiously, which will have a red colour, into a glass flask. Then pour fresh vinegar onto it, and put it into Digestion as before, so that that which may have remained with the faecibus, it should thus have ample time to become dissolved. Then the faeces may be discarded, for they are no longer useful, except for being scattered over the earth and thrown away.

Afterwards pour all the solutions together into a glass retort, put into Balneum Mariae, and distill the sharp vinegar off and pour it back on again. Or, rather a fresh one, since the former would be too weak, and the matter will very quickly become dissolved by the vinegar. Distill it off again, so that the matter remains quite dry. Then take common distilled water and wash away all sharpness, which has remained with the matter from the vinegar, and then dry the matter in the sun, or otherwise by a gentle fire, so that it becomes well dried. It will then be fair to behold, and have a bright red color.

The Philosophers, when they have thus prepared our Antimonium in secret, have remarked how its outermost nature and power has collapsed into its interior, and its interior thrown out and has now become an oil that lies hidden in its innermost and depth, well prepared and ready. And henceforth it cannot, unto the last judgement, be brought back to its first essence. And this is true, for it has become so subtle and volatile, that as soon as it senses the power of fire, it flies away as a smoke with all its parts because of its volatility.

Several poor and common Laborers, when they have prepared the Antimonium thus, have taken one part out, to take care of their expenses, so that they may more easily do the rest of the work and complete it. They then mixed it with one part Salmiac, one part Vitro (alii. Nitro, alii. Titro), one part Rebohat, to cleanse the Corpera, and then proceeded to project this mixture onto a pure Lunam. And if the Luna was one Mark, they found two and a half Loth good gold after separation; sometimes even more. And there with they had accomplished a work providing for their expenses, so that they might even better expect to attain to the Great Work. And the foolish called this a bringing into the Lunam, but they are mistaken. For such gold is not brought in by the Spiritibus (alii. Speciebus), but any Luna contains two Mark gold to the Loth, some even more. But this gold is united to the Lunar nature to such a degree that it may not be separated from it, neither by Aquafort, nor by common Antimonium, as the goldsmiths know. When however the just mentioned mixture is thrown onto the Lunam in flux, then such a separation takes place that the Luna quite readily gives away her implanted gold either in Aquafort or in Aqua Regis, and lets herself separate from it, strikes it to the ground and precipitates it, which would or might otherwise not happen. Therefore it is not a bringing into the Lunam, but a bringing out of the Luna.

But we are coming back to our Proposito and purpose of our work, for we wish to have the Oil, which has only been known and been acquainted with this magistry, and not by the foolish. When you then have the Antimonium well rubified according to the above given teaching, then you shall take a well rectified Spiritum vini, and pour it over the red powder of Antimony, put it in a gentle Balneum Mariae to dissolve for four days and nights, so that everything becomes well dissolved. If however something should remain behind, you overpour the same with fresh Spiritu vini, and put it into the Balneum Mariae again, as said before, and everything should become well dissolved. And in case there are some more faeces there, but there should be very little, do them away, for they are not useful for anything.

The Solutions put into a glass retort, lute on a helm and connect it to a receiver, also well luted, to receive the Spiritus. Put it into Balneum Mariae. Thereafter you begin, in the Name of God, to distill very leisurely at a gentle heat, until all the Spiritus Vini has come over. You then pour the same Spiritum that you have drawn off, back onto the dry matter, and distill it over again as before.

And this pouring on and distilling off again, you continue so often until you see the Spiritum vini ascend and go over the helm in all kinds of colours. Then it is time to follow up with a strong fire, and a noble blood red Oleum will ascend, go through the tube of the helm and drip into the recipient. Truly, this is the most secret way of the Wise to distill the very highly praised oil of Antimonii, and it is a noble, powerful, fragrant oil of great virtue, as you will hear below in the following.

But here I wish to teach and instruct you who are poor and without means to expect the Great Work in another manner; not the way the ancients did it by separating the gold from the Luna. Therefore take this oil, one lot[22], eight lot of Saturn calcined according to art, and carefully imbibe the oil, drop by drop, while continuously stirring the calx Saturni. Then put it ten days and nights in the heat, in the furnace of secrets, and let the fire that this furnace contains, increase every other day by one degree. The first two days you give it the first degree of fire, the second two days you give it the second degree, and after four days and nights you put it into the third degree of fire and let it remain there for three days and nights. After these three days you open the window of the fourth degree, for which likewise three days and nights should be sufficient. Then take it out, and the top of the Saturnus becomes very beautiful and of a reddish yellow colour. This should be melted with Venetian Boreas. When this has been done, you will find that the power of our oil has changed it to good gold. Thus you will again have subsistence, so that you may better expect the Great Work.

We now come back to our purpose where we left it earlier. Above you have heard, and have been told to distill the Spiritum vini with the Oleum Antimonii over the helm into the recipient as well as the work of changing the Saturnum into gold. But

---

[22] A LOT is an ancient unit of measurement primarily used in weighing gold and silver. The modern equivalency is about 1/30 of a pound.

now we wish to make haste and report about the second tinctural work. Here it will be necessary to separate the Spiritum vini from the oil again, and you shall know that it is done thus: Take the mixture of oil and wine spirit, put it into a retort, put on a helm, connect a receiver and place it all together into the Balneum Mariae. Then distill all the Spiritum vini from the oil, at a very gentle heat, until you are certain that no more Spiritus vini is to be found within this very precious oil. And this will be easy to check; for when you see several drops of Spiritu vini ascend over the helm and fall into the recipient, this is the sign that the Spiritus vini has become separated from the oil. Then remove the fire from the Balneo, though it was very small, so that it may cool all the sooner.

Now remove the recipient containing the Spiritu vini, and keep it in a safe place, for it is full of Spiritus which it has extracted from the oil and retained. It also contains admirable virtues, as you will hear hereafter.

But in the Balneo you will find the blessed bloodred Oleum Antimonii in the retort, which should be taken out very carefully. The helm must be very slowly removed, taking care to soften and wash off the Lute, so that no dirt falls down into the beautiful red oil and makes it turbid. This oil you must store with all possible precaution so that it receives no damage. For you now have a Heavenly Oil that shines on a dark night and emits light as from a glowing coal. And the reason for this is that its innermost power and soul has become thrown out unto the outermost, and the hidden soul is now revealed and shines through the pure body as a light through a lantern: Just as on Judgement Day our present invisible and internal souls will manifest through our clarified bodies, that in this life are impure and dark, but the soul will then be revealed and seen unto the outermost of the body, and will shine as the bright sun.

Thus you now have two separate things: Both the Spirit of Wine full of force and wonder in the arts of the human body: And then the blessed red, noble, heavenly Oleum Antimonii, to translate all diseases of the imperfect metals to the Perfection of gold. And the power of the Spiritual Wine reaches very far and to great heights. For when it is rightly used according to the Art of Medicine: I tell you, you have a heavenly medicine to prevent and to cure all kinds of diseases and ailments of the human body. And its uses are thus, as follows:

AGAINST PODAGRA OR GOUT

In the case of gout one should let three drops of this Spiritu vini, which has received the power of the Antimony, fall into a small glass of wine. This has to be taken by the patient on an empty stomach at the very moment in time when he senses the beginning or arrival of his trouble, bodily ailment and pain. On the next day and afterwards on the third day it should also be taken and used in the same way. On the first day it takes away all pain, however great it may be, and prevents swelling. On the second day it causes a sweat that is very inconstant, viscous and thick, that smells and tastes quite sour and offensive, and occurs mostly where the joints and limbs are attached. On the third day, regardless of whether any medicine has been taken, a purging takes place of the veins into the bowels, without any inconvenience, pain or grief. And this demonstrates a great power of Nature.

AGAINST LEPROSY

To begin with the patient is given six drops on an empty stomach. And arrange it so that the unclean person is alone without the company of any healthy people, in a separate and convenient place. For his whole body will soon begin to smoke and steam with a stinking mist or vapor. And on the second day his skin will start to flake and much uncleanliness will detach itself from his body. He should then have three more drops of the medicine ready, which he should take and use in solitude on the fourth day. Then on the eighth or ninth day, by means of this medicine and through the bestowal of Divine mercy and blessing, he will be completely cleansed and his health restored.

AGAINST APOPLEXIA OR STROKE

In the case of stroke, let a drop of the unadmixed tincture fall onto the tongue of the person in need. At once it will raise itself and distribute itself like a mist or smoke, and rectify and dissolve the struck part. But if the stroke has hit the body or other members, he should be given three drops at the same time in a glass of good wine, as previously taught in the case of Podagra.

AGAINST HYDROPE OR DROPSY

In the case of dropsy give one drop each day for six days in a row, in Aqua Melissae or Valerianae. On the seventh day give three drops in good wine. Then it is enough.

AGAINST EPILEPSIA, CATALEPSIA and ANALEPSIA

In case of the falling sickness, give him two drops at the beginning of the Paroxismi in Aqua Salviae, and after three hours again two drops. This will suffice. But if further symptoms should occur, then give him two more drops as above.

AGAINST HECTIC

In case of consumption and dehydration, give him two drops in Aqua Violarum the first day. On the second day, give him two more drops in good wine.

AGAINST FEVER

In cases of all kinds of hot fevers, give him three drops in a well distilled St. Johnswort water or Cichorii at the beginning of the Paroxismi. Early in the morning on the following day, again give him three drops in good wine on an empty stomach.

AGAINST PEST

In the case of pestilence give the patient seven drops in a good wine, and see to it that the infected person is all by himself, and caused to sweat. Then this poison will, with Divine assistance, do him no harm.

FOR THE PROLONGATION AND MAINTENANCE OF A HEALTHY LIFE

Take and give at the beginning and entry of spring, when the sun has entered the sign of Aries, two drops; and at the beginning with God's help, be safe and protected against bad health and poisoned air, unless the incurred disease was predestined and fatally imposed upon man by the Almighty God.

But we now wish to proceed to the Oleum Antimonii and its Power, and show how this oil may also help the diseased and imperfect metallic bodies.

Take in the Name of God, very pure refined gold, as much as you want and think will suffice. Dissolve it in a rectified Wine, prepared the way one usually makes Aquam Vitae. And after the gold has become dissolved, let it digest for a month. Then put it into a Balneum, and distill off the spiritum vini very slowly and gently. Repeat this several times, as long and as often until you see that your gold remains behind in fundo as a sap. And such is the manner and opinion of several of the ancients on how to prepare the gold.

But I will show and teach you a much shorter, better and more useful way. Viz. That you instead of such prepared gold take one part Mercurii Solis, the preparation of which I have already taught in another place by its proper process. Draw off its airy water so that it becomes a subtle dust and calx. Then take two parts of our blessed oil, and pour the oil very slowly, drop by drop onto the dust of the Mercurii Solis, until everything has become absorbed.

Put it in a vial, well-sealed, in to a heat of the first degree of the oven of secrets, and let it remain there for ten days and nights. You will then see your powder and oil quite dry, such that it has become a single piece of dust of a blackish, grey colour.

After ten days give it the second degree of heat, and the grey and black colour will slowly change into a whiteness so that it becomes more or less white. And at the end of these ten days, the matter will take on a beautiful rose white. But this may be ignored. For this colour is only due to the

Mercurio Solis, that has swallowed up our blessed oil, and now covers it with the innermost part of its body. But by the power of the fire, our oil will again subdue such Mercurium Solis, and throw it into its innermost. And the oil with its very bright red colour will rule over it and remain on the outside.

Therefore it is time, when twenty years (sic) have passed, that you open the window of the third degree (The alchemical oven had openings or 'registers' by which the heat could be controlled.) The external white colour and force will then completely recede inwardly, and the internal red colour will, by the force of the fire, become external. Keep also this degree of fire for ten days, without increase or decrease. You will then see your Powder, that was previously white, now become very red. But for the time being this redness may be ignored (is of no consequence), for it is still unfixed and volatile; and at the end of these ten days, when the thirtieth day has passed, you should open the last window of the fourth degree of fire, Let it stay in this degree for another ten days, and this very bright red powder will begin to melt. Let it stay in flux for these ten days. And when you take it out you will find on the bottom a very bright red and transparent stone, ruby colored, melted into the shape of the vial. This stone may be used for Projection, as has been taught in the tract on Vitriol. Praise God in Eternity for this His high revelation, and thank Him in Eternity. Amen.

ON THE MULTIPLICATION - LAPIDIS STIBII

The ancient sages, after they had discovered this stone and prepared it to perfect power and translation of the imperfect metals to gold, long sought to discover a way to increase the power and efficiency of this stone. And they found two ways to multiply it: One is a multiplication of its power, such that the stone may be brought much further in its power of Transmutation. And this multiplication is very subtle, the description of which may be found in the Tract on Gold.

The second multiplication is an Augmentum quantitatis of the stone with its former power, in such a way that it neither loses any of its power, nor gains any, but in such a manner that its weight increases and keeps on increasing ever more, so that a single ounce grows and increases to many ounces.

To achieve this increase or Multiplication one has to proceed in the following manner: Take in the Name of God, your stone, and grind it to a subtle powder, and add as much Mercurii Solis as was taught before. Put these together into a round vial, seal with sigillo Hermetis, and put it into the former oven exactly as taught, except that the time has to be shorter and less now. For where you previously used ten (alii thirty) days, you may now not use more than four (alii ten) days. In other respects the work is exactly the same as before.

Praise and thank God the Almighty for His high revelation, and diligently continue your prayers for His Almighty Mercy and Divine blessings of this Work and Art as well as His granting you a good health and fortuitous welfare. And moreover, take care always to help and counsel the poor.

LAUS DEO OMIVPOTENTA

NOTA. He who wishes to know more about Antimonio may consult Fr. Basilii Valentini, *"Triumphal Chariot of Antimony"*[23] with comments by Theodor Kerckringius.

---

[23] Volume 2 in the R.A.M.S. Library of Alchemy.

The Truest and Most Noble Secret

The Reverend Teacher and Presbyter:

Joducus Greverus

Extracted from Theatrum Chemicum, Vol. III, Page 699

Translated from Latin by Patricia Tahil

To the honour of the holy and undivided
Trinity, and of blessed Mother Mary, ever virgin, to
console the pious, help the poor, and direct
philosophers, I, Jody Grever, presbyter and least of
Philosophers, with great error, expense and labour
and passing of my life in the quest, not despairing
of Divine grace and mercy, strengthened by the help
of God himself start to describe to you the true and
perfect work, not in parts, nor in ambiguous and
obscure words, for He who alone has the power to
pass on and impart this most secret art to whom He
will, has given it to me of his goodness.

Therefore, to the glory of Him who has given so
much to me, an unworthy sinner, I celebrate and
proclaim with all my power those whom he has called
and chosen to these wonderful mysteries of
Philosophy, and I shall make the matter plain in
proper order, without learned obfuscation, in speech

clear and open but nevertheless confirming to the methods and customs of the Philosophers.

I know well enough that no one can seize this knowledge for himself unless God confers it on him. Who alone makes plain the darkness of this mystery, or darkens its clarity, for none can understand the plainest things if not illuminated by God, nor can he follow the most obscure if God does not lighten them. However, I adjure all of you who have my writings, not to communicate them to the unworthy, or the greedy, or those who are unkind to their neighbours, or oppress them, such as tyrannous men, the unjust, adulterers, the weak and those whose God is their belly. Place your trust, therefore, in the Lord God Creator of all and work in fear of him, using what you have to hand according to what I have written, and awaiting the blessing of our heavenly Lord.

First of all, O best beloved, you should know that our art imitates nature; and where nature produces plants and trees by herself, which the gardener then increases, by sowing their seeds, or grafting their twigs, so she generates metals in the depths of the mountains of the earth, which the artificer increases by grafting or sowing them in the proper soil or a suitable stock. What you must know above all is what the seeds of gold and silver are, what their soil, and what their stock. We tell you. Therefore, that Sol and Luna contain their own seeds in themselves, that Saturn, Jupiter, Mars and Venus are their stocks, and the proper soil for all of them is Mercury; it is the stable foundation to which you should direct your whole attention, accommodating the appearance of the one so that is suitable for the nature of the other, because God the author of nature wishes us to understand and perfect this area of philosophy by this sort of contemplation; by means of these writings we communicate it to the worthy and the chosen, so that I and others with me may offer perpetual praise and glory to God, giving due thanks for his grace.

Just as the Philosophers, therefore, when contemplating the works of nature, are accustomed to seek the likeness of the Stone in plants, animals, trees, eggs, foetuses, the joining of masculine and feminine and an almost infinite number of the same sort of things, so we too pave the way to a clearer understanding of this particular operation, we follow nature's tracks and choose thence the true and necessary adaptations.

However, even if this is a special operation, nevertheless there is truth in it, so that anyone who looks deeply into the causes of things, is close to observation of the great triumphal Stone in certain truth and power. By the aforementioned simile I have already indeed shown the wise what is the material that should be taken to start the work with; and that I may further uncover it to your eyes, I say that just as a man comes from a man, an ass from an ass, a hen from a hen's egg, and corn from corn, so everything is made from its likeness, gold from gold, and silver from silver.

You will therefore understand from this what material you should take and bring into the work; here you should also know that a man cannot always be produced from any man, nor a dog from any dog nor a plant from any plant of its own species; for a boy of twelve years old, or a person of either sex who is languishing in sickness will not bring forth a human fetus from their mutual intercourse; neither will a pup a dog, nor young or rotten plants a (new) plant. But a mature and healthy man and a woman who is similarly mature and strong will procreate a human fetus; likewise other animals after their own kind and ripe seed from mature plants. Therefore we say that as farmers carefully select the seeds of fruits, plants and trees, and do not take them unless they are ripe, and do the same when they take the strongest shoots from the trees for grafting, so in this way you will be able to judge what you take for this business.

For common gold does not enter into the work, nor common Mercury, nor vulgar silver, nor any other common thing, but rather philosophical things; what you may take these to be is clearly shown from the aforesaid. However, what has been said of the seeds must also be said of the soil, tor not all soils are suitable for this or that plant, as Virgil himself warns us in the Georgics, and as we see farmers and gardeners always observing. You must therefore look, O best beloved, and see that the soil is suitable for growing the seeds of Sol or Luna, for if you choose foolishly, you will miserably lose your seed, your labours and your expense.

Seeds and soil must therefore be chosen with great care, and the chosen soil must be cultivated in the same way as the earth of the fields and gardens is cultivated, with ploughing or digging, manuring, harrowing, watering, and all that sort of thing. Geber and the workers in this divine task call choosing and culturing the seeds and the soil preparation, and without it one cannot have either Sol or Luna of the Philosophers, nor can one find a clear way in to the secrets of this great art.

Moreover, common gold is very impure, contaminated by admixture with other metals, sick, diseased, and therefore sterile, and so is common silver, while on the contrary Sol and Luna of the

Philosophers are very pure, not besmirched by the addition of any foreign mixture, healthy, strong and abounding in reproductive seed. Common earth is also uncultivated and unploughed, but the Philosophers' earth is cultivated by fire-breathing bulls, as it says in the fable of the Golden Fleece.

And so, O best beloved, find out the true meaning of my words, and understand that the Philosophers are like farmers and gardeners, who first choose certain seeds and having selected them do not sow them in common earth, nor in uncultivated fields, but in garden ridges, and once they have sown them they commit them to the heat of the Sun and the goodness of earth and sky, and await their due fruition after a suitable time. The Philosophers employ similar reasoning, for before they begin to work they must decide to take in hand vulgar gold and silver, whether sick or healthy; for if they are found to be sick, the right medicine will render them healthy as necessary, which is the preparation itself. This doctored gold, healthy and purified, is the Philosophers' Sol, and the same is to be understood of Luna.

However, when you have philosophical Sol and Luna, like good seeds the soil must be cleansed of its dirt and useless weeds, and improved by careful cultivation; and in this improved earth you must sow the aforesaid seed of Sol and Luna, and commit them to continuous heat and the goodness of earth and sky; nothing radical will happen before the right time, but you must wait for the ripe fruits after the leaves and flowers.

From this it is clear that even if the Philosopher takes vulgar gold, silver, or Mercury, he does not bring them into the work until he has raised them from common earth to the grade of a medicine. There are some who wrap these things up in great obscurity, and throw the Philosopher's soul into great doubt, all of which you will be able to resolve if you observe the general rule that I give: nothing must enter the medicated work without the proper preparation, whose ability is to make medicines of common substances. And so nothing filthy, or diseased, or uncultivated should enter our work; before the work itself we may take the sick, the filthy and the uncultivated.

So the Philosopher should take the material given by nature, prepare it properly, choose and cultivate the proper soil, then put in the seed and apply the correct continuous heat. It is, however, Nature's part to rot and dissolve the seed placed in the ground, make roots, draw up the dryness of earth, produce shoots, leaves and flowers in succession, conceive, perfect and ripen the seed,

Having reached this stage nature comes to a stop, and goes no further unless stirred by the Philosopher, who must know how to make nature progress by certain correct signs, not by false opinions.

Now when he sees that the fruit is exactly ripe, he must not delay, but at once collect the mature seeds of the fruit, separated from any excess matter they may have and immediately fresh soil must be cultivated, and when it has been well tilled, he puts the seeds in it, and applies continuous heat as before, whereupon nature again takes up her task, finishes the work and little by little multiplies perennial fruit and seed.

From this consideration it arises as a conclusion that natural things require different lengths of time to complete their generations, for certain plants grow and mature in about a year. Some trees mature more quickly, others more slowly. So it is with metals, some ripen quickly, others are slower.

Furthermore, just as we see farmers at once disturb and cover the seeds by harrowing, and bind the grafts inserted in stocks with wax or clay, which are not tree-like by nature, but are helpful intermediaries, (making sure) that the sap of the stock enters the graft only, so in this work a certain clay or wax must be taken that forbids the juice of the earth to exhale, and contains it within; moreover farmers customarily bind this wax with straw so that the onslaught of rain and wind may not tear out the inserted graft from the stock at the joint. So we do in our work such things as we observe in the everyday world, and if you consider this carefully, you will be easily freed from many errors, and pursue the work to its desired end.

But let us dally no more, but come to the practice in its proper order; and let us first speak of the tools and materials requisite for this work, then of the preparations, next of the seeds and the ripening, and finally of the repetition of the sowing, which is the complement of the work.

THE TOOLS FOR THIS TASK

The tools required for this task are as follows: 1. A subliming furnace. 2. A digesting furnace which can also be an athanor and be used for ashes or sand. 3. A cupel or jar for ashes or sand. 4. Sieved ashes or sand. 5. A subliming jar. 6. Six or seven small subliming aludels. 7. Clay shaken with horse-dung. 8. An oblong piece of iron curved at one end into a spatula. 9. Bellows. 10. Tongs for lighted coals. 11. Coals. 12. A round glass cucurbit. 13. Some crucibles. 14. Cupels or shallow dishes made of bone ash. 15. An anvil. 16. A hammer.

The materials are lead, antimony, gold, silver, bronze, and quicksilver. Also yellow wax, vitriol, sal nitrum, alum, quicklime, bronze, glass, and a urinal with a wooden spatula. And so the total number of tools and materials comprises the number four times seven, which is a mystery. The athanor furnace is almost the same as the subliming furnace, except that a cupel containing ashes or sand is placed on it, and in this is placed the round glass cucurbit with the material, and it is closed from above with an empty cupel, so that no draught nor air can touch the glass.

PREPARATION OF THE MATER ALS: GOLD FIRST

Choose fresh gold for this task that has not been used by the craftsman for many operations, but has come recently from the mine and has not been much worked, of the best color, about twenty-four carats, or nearly; and the best of this sort is that which is found in small pieces in the ground, as if cooked and semi-liquified. If you can have this, no other preparation for the work is necessary beyond hammering it into the thinnest flakes.

But if you cannot get it, take gold from Hungarian florins, and purge it, or have it purged by melting it into a regulus with antimony, as the workers who make gold-leaf do, for antimony has the property of purifying gold from all superfluities.

If perchance you do not know the purification by antimony, you can cement it by heating for twenty-four hours, reduce it to a mass, and hammer it into the thinnest flakes.

In order to give you a complete account, however, we shall describe to you the method of purifying gold with antimony, which is as follows: Take one ounce of gold, or any reasonable quantity, and five times as much antimony as gold, and put the gold and the antimony in a crucible, and place it on burning coals, and allow it to flux for an hour, or half an hour, then take an implement of iron or brass shaped like a pyramid and hollow within, very smooth inside and heated by the fire, and grease the interior with wax or a candle and after you have taken the gold and the antimony off the heat, pour it from the crucible into this funnel, and by striking the bottom of the funnel gently, the gold will fall down separately. Then allow it to cool and invert the funnel, and take out the pyramidal mass from the funnel. Its apex will contain the regulus of gold, which is separated from the rest of the material by hitting it just below the apex, and keep the regulus of gold aside.

Now take the mass of antimony and add to it one third of its weight of fresh antimony and liquify in the crucible again and pour again into the funnel, separate this regulus in its turn, if there is any, and add it to the previously separated regulus. You can do it a third time, if necessary, or not, as you will know from the amount of second regulus (obtained); if it be very little, leave it. Keep this aside also, and proceed with the regulus in this manner:

Take a piece of tile made of tile-bricks, like those that used to be on the roofs of the towers of ancient churches, and make a small hole in it to hold your reguli. Place it in a dry spot, and surround it with coals, putting the reguli in the hole, then light the coals and allow the antimony to exhale from the reguli in smoke; when the fuming ceases, cover it carefully with coals, and blow on it from above with a bellows until you can see no spot or speck on your gold, but it is beautiful and greens like grass in the fire; this usually happens in two or three hours.

Having then taken out the piece of brick with the gold that was purged in it, quench it in clear urine, and so your gold will be perfectly cleansed, and ready, so that it can be hammered into the finest sheets, which are to be kept wrapped in paper till the proper time. If you can find a sufficient quantity of fine gold-leaf among the druggists or painters, that may be used as pure gold, needing no further preparation. If they have enough Pure silver-leaf, do the same; if not, purify it as follows.

PREPARATION OF SILVER

Like gold, the best silver has come fresh from the mine, and has not been much worked by the craftsman, but has been cupelled with lead in a dish or cupel. If, however, you do not have such silver, take what you have, and if you suspect that copper is mixed with it, or any other impurity, purify in the cupel with two or three times its weight of black lead, until it is cleansed of all impurity. Then make fine sheets of it, and, keep pure in clean paper until it is needed.

PREPARATION OF MERCURY

You should choose argent vive which has not been made by artifice, from lead or any other metal, but has come from its own mineral, such as mountain cinnabar or marcasite. However, you should prepare such mercury for our work like this:

Take a pound of it, three ounces of vitriol, three ounces of quick lime, yellow wax one pound, and grind the vitriol, sal nitrum, alum, and quicklime in a mortar and mix them. Then place the wax in a glass urinal, the bottom of which has been luted with clay mixed and sieved with horse dung,

265

and put the urinal in a dish containing sand on a small furnace, and subject it to heat, gentle at first, so that the glass does not break from sudden heat, until the wax melts.

Meantime, shake up the mercury in a glass dish with the aforesaid substances, in order to mix them thoroughly, having poured in a little acid from strong wine. When the wax in the urinal has melted, add the mercury with the aforementioned powders little, and heat in the wax for three hours over a medium heat, constantly stirring and turning with a wooden spatula.

When the three hours are up, let it cool in the urinal, and make a hole in the wax with a clean rod before it hardens, and this is how you keep your Mercury, well prepared for this work and ready to have Sol and Luna sown in it.

PREPARATION OF ANTIMONY

(The Uniting Medium)

Prepare the antimony, which enters into this work as a tie between two substances, by subliming in this way:

Take a pound of it, and grind to a powder in a mortar, then place in a subliming jar, and place upon it an aludel with a hole in the side; this aludel must be open both at the wider part and at the narrower, and so must the rest of the aludels.

Now apply firm clay mixed with horse dung to the joints, and join all six aludels in a row like this, so that the lower narrower part enters the wider upper part from beneath, all the joints having been firmly luted. However, the last aludel should have a smaller hole than the others, as big as a finger, no wider, that can easily be closed by a dry wooden plug.

When the joints have dried, put this subliming tower on your furnace, and secure it with a strong lute, so that it does not wobble to and fro, and leave four holes around it to let out the smoke from the furnace, then light the coals in the furnace, and stoke up the fire by stages, until the powdered antimony liquefies; this you will know for certain if you put an iron spatula obliquely into the subliming jar through the lateral hole in the lowest aludel.

When you feel that the powdered antimony in the jar has liquefied, then constantly turn and move the antimony with your iron spatula, applying a strong fire beneath it, until the bottom of the sublimatory jar that is in the furnace reddens. Continue the heating and the stirring of the antimony until you feel that all the antimony has sublimed and no more is coming off.

You can be sure of this if upon opening the stopper of the topmost aludel you see no fumes exhaling from the opening. Then reduce the heat gradually and carefully put a moist cloth round the joint of the lowest aludel, and when the joint has absorbed the moisture remove the lute from all the aludels carefully, and collect the white sulphur of antimony from them which has sublimed on to their sides like a white flour. This sulphur is wonderfully effective in our work, for it is the friend of Mercury, Sol and Luna, the bond of body and spirit, the marriage of spirit and soul; it is, I say, the vehicle of the tincture and the wax that holds in the vital sap of the stock, so that it flows only into the graft you have inserted into it.

Collect this sulphur carefully, therefore, and keep it like precious treasurer by its use you will attain the desired end of your work. And so we complete our discourse on the preparation of all the materials necessary and sufficient for this precious secret work, nor have we altered the order of what we have said. Now it remains to speak of the sowing.

PUTTING THE SEEDS IN THE GROUND

Beloved worker, you have already completed a great part of your task; it remains for you to prepare your ground for putrefaction, so that it may give the desired fruit, twofold, threefold, a hundredfold and a thousandfold, for unless a grain of corn is thrown into the soil and dies, it will bear no fruit.

In the name of the holy and undivided Trinity, therefore, take an ounce of your gold leaf, or whatever reasonable quantity you wish, and four times that amount of the well-cultivated earth that is in the glass urinal, covered by a bee-hive. If, however, you are Poor and cannot obtain so much gold, then take the weight mentioned of silver-leaf and six times the amount of cultivated earth.

So divide whichever luminary you have into pieces, and add it to the aforesaid earth in a cold crucible, stirring continually with your fingers, or a wooden or glass pestle until all the flakes are mixed with our earth. Then have ready another clean crucible, put it on burning coals, and when it begins to glow take it off the fire and place it in a safe place between two or three bricks, so that it cannot easily be overturned, and have ready a clean glass dish with pure water in it, and then throw your flakes mingled with earth on to the hot crucible. Stop up your mouth so as not to breathe the fumes of the earth, and stir the flakes with a wooden stick, so that they enter the earth, which usually occurs quickly, and when the earth has taken them up, which you will know because there are no more lumps or globules under your stick, immediately pour it into the dish of pure cord water, and thus you have sown your seed in your earth.

The Philosophers commonly call this mixture "amalgam", which is the same as saying "softened", because the hard seed of gold or silver is softened in your earth. This mixture must, however, be washed and rubbed with the purest $tater and the whitest purest salt until no further blackness appears in it, and the water that enters and leaves it is clear. Then dry the moisture adhering to it with a clean cloth. This then is the beginning of the work, which philosophers have unworthily tried to conceal by calling it by many different names, such as our bronze, earth of magnesia, the whole composition, etc. Now we proceed to the harrowing.

HARROWING THE SOWN FIELD

You must recognize, O best beloved, that this part of the operation is called impregnation by some Philosophers, and the perpetual disposition by others; but we do not fear to call it harrowing, following our simile at the beginning. For just as the sown field is turned over by harrowing, and its larger clods broken up, so that the seeds which have been sown in it are covered, so in our work, our Mercurial earth, which has accepted the seeds of whichever luminary it may be, is ground and turned over, and the seeds are wholly hidden in it.

So when you have sown your field, that is to say, made your amalgam, and washed and dried it, then take the white flour of sublimed antimony, and sprinkle a little of it on the amalgam on a flat piece of marble, and grind the whole until the said amalgam has taken up the white antimony. When you have done this, add more of the said white antimony, and again grind them together until it takes up no more of the white sulphur of antimony. This is the sign intended to show that the harrowing is finished.

Collect your amalgam now ready for impregnation, and carefully place it in a round glass vessel, which is large enough for your sown and harrowed earth to occupy one third of it, leaving two thirds empty. Then close the opening of your round vessel with a clever lute made from white of egg, brick dust and quicklime, spread on a cloth moistened with egg-white. Allow it to dry, and you will have the perfect Philosophers' Egg, which fulfills the saying of the Philosophers. In one vessel only, one stone, one digestion. But this belongs to the business of ripening, of which it is now time to speak.

GROWTH OF THE SEED

Philosophers employing the simile of an egg, as we did that of the sowing of plants, taught that the first state of the egg was putrefaction, and this they extended, not only to the formation of the chick in the egg, but to its hatching; and appended feeding to hatching. We, employing a simile that is clearer to the intellect, speak of these things differently, like this: We see that the seeds of plants do not rot entirely, nor are they completely mixed with earth, but softened and moistened by the wet juice of the earth, first they put out a shoot, then a tuft grows, and umbels come forth, soon it flowers, then sets seed, which finally ripens; so we are of the opinion that the same stages must be observed in this task, and we advise you to adjust the heat each month, so to say.

For when you have closed the Philosophic egg, and you can still see the snowy whiteness of the earth over everything, I warn you that you ought to use the gentlest heat of March. Now by the application of this heat, that snowy white turns into a greyish or cinder color, which indicates that the shoot risen from the earth has begun to grow, and that April's influence has arrived. Then the risen shoot grows by degrees in May's heat and power, and leaves and twigs increase together.

Then afterwards with the heat of June it puts forth shoots and umbels and swells them; from these come flowers in burning July. In the heat of August it forms and joins up the seeds, which are ripened partly in the same month, but mostly by the heat present in September.

I am sure that the regulation of the heat is plain to any diligent man from what I have already said, but so that the crasser may also be catered for, we make it clear by laying it, out in order a little more openly.

When you have closed your Philosophical egg, you will have the oven of your athanor ready with a dish of glass and copper, about six or seven finger-breadths deep. In this put a layer of fresh horse-dung or softest hay, two fingers deep, or a little more, upon which you are to place your glass egg. Pack it round with dung or hay carefully, so that it cannot move, but do not pack any higher than the depth of the material in the round vessel.

If you observe that the glass does not stand firmly enough, do not be ashamed to make a circular piece from a plank, and to make a hole in it of the size that will best fit the glass. In another part of it, away from the middle, make another finger-sized hole, through which you will be able to pour water as often as necessary. Then put the plank on the dung surrounding the round vessel, so that part of the vessel appears through the middle of the hole, and place two or three pieces of brick on the plank, so that it is held firmly.

When you have done this, pour in as much clear water as will cover the surface of the plank no deeper than the thickness of a reed. Then close your oven with the closure of the athanor, and put a gentle heat underneath that warms as much as the water-bath or horse-dung, as much as feels like the touch of a hot needle, but always without boiling. This you must watch carefully, frequently removing and replacing the cover of the athanor.

Continue this heat of early Spring, or the first grade, or the month of March, for seven days and nights, taking the precaution that when you see the water evaporate with the continual and gentle heat you immediately put a funnel in the hole in the plank, and again pour in warm water, or even cold water, so that the water does not begin to boil from neglect.

When the required number of days has elapsed, take out the round vessel, and see whether the original color has yet turned from white to dark cinder color, or black diluted with white, and put it back in its place before it gets cold. Be careful that your hand is not very cold when it touches the vessel, otherwise it will break.

If you see the color I have spoken of, rejoice, for you have supplied the correct color, and you are looking at the true sign that the seed has already germinated. Do you not see that all shoots are yellow when they first burst from the earth? You must also know, direst investigator of so great a secret, that in this work nothing equals observing with great care the color and the amount of moisture; these are the two things that lead the worker without error from the beginning to the end of the task.

It is for this reason that Geber orders the worker carefully to fix in his mind the peculiarities of each digestion, and to inquire what causes them. But this will not help the stiff-necked worker, nor the greedy, nor the weak, nor any who is enmeshed in the vanities of this world. If the sign does not appear before your eyes, continue the same degree of heating for four more ordinary days, and then look at it again, and replace it again if the signs do not yet appear, and do this every four days until the signs we have told you of do appear.

During this digestion a certain mercury-like vapour ascends from the earth like a cloud and will stick to the sides of the upper empty part of your egg. This you are to leave intact until you have learnt from its meaning what to do next. Do you not see a sown field giving off many and various vapours both before and after germination? (Do you not see) how some of them are turned into rain and water the field again and supply useful nourishment to the shoots, while others disappear into the air and carry nothing useful to the shoots? Consider carefully, therefore, what is to be done here, and do not rashly convert these exhalations into rain that turns back of its own accord and waters your field.

When you have plain evidence that your seeds have germinated you will know that March has gone and that it is the beginning of the mild Spring of April, and that you should replace the color that hardly differs from that of March with great care, without boiling the water on the dung, so that the shoot born of the seed begins to grow, and strengthen. The sign that the shoot is sprouting is a more plentiful exhalation of moisture, and a change in the aforesaid color into a dark one, with a scarcely noticeable redness. You are to expect the appearance of this sign on the seventh or eighth day after you have begun to apply the heat of early

Spring. And you are to maintain such a regimen during these two intervals of time that the germination of the seed by putrefaction and the sprouting of the shoot by application of a digestive heat are both accomplished.

GROWTH OF THE SHOOT

With the end of early Spring, we cone to late Spring, when the shoot has already become a plant, begins to raise its twigs, and shows its proper green color. You are therefore to imitate May-time by applying the greatest heat that water is capable of, without boiling, however, but with the shiver of boiling; this you are to continue until the following signs appear.

First the material shows a dull reddish color on the side, almost black, neither black, nor red, nor brown, but what you will recognize as a mixture of all: these. Secondly there is the dryness of the earth, from which you will see if you have properly judged the quantity of the previous exhalations. For if you have clearly seen that there are fewer exhalations with this heat, and that those which are already adhering to the sides of your vessel are not increasing, but remain the same in quantity, you may judge that your earth is dry.

Upon seeing the signs mentioned, you will immediately understand that your shoot has developed and grown and become firm with the hardness proper for grass or reeds. Remember however, son, in case the Philosophers' method of speaking should deceive your that we are not describing multiplication, but the development of the shoot; for there is no multiplication of the shoot until after the multiplied seed has been sown afresh. Development of the shoot occurs before the seed can multiply in it. Have you not considered the progress of all natural things?

Where have you seen the individual member of any species multiply before its seed has germinated, sprouted, and divided into its parts? And put forth stems, and umbels or clusters, which flower at length, and seeds and fruit are formed from the flowers, and when formed ripen under the influence of the right heat? Are not these many seeds formed from one seed, and when they have ripened properly must you not collect them, cleanse then of their husks, remove their chaff, take them from their shells, and then sow them again, and wait for ripening through the same steps of growth as nature put them through before, and then do you not expect the first multiplication, and finally see and have the first multiplication of the individual?

O fools, what madness! Will you dismiss reason and the works of nature and follow the irrational fantasies of your brain? Does not Geber tell us that all the actions of nature have their predetermined time, and that that which produces an effect is limited to a greater or lesser extent? But indeed, if you are so lacking in reason and intelligence that you cannot understand these things from nature herself and her workings, leave off this most secret art, and flee to the countryman's plough handle and plough, or at least learn from them by imitating nature's ways.

For you indeed, my son, initiated into the study of Philosophy, have already long understood our words where we have not written clearly or intelligibly enough; we have hidden nothing but two things, which no philosopher has ever dared to write openly of, because they are only passed on privately to sons of the doctrine. And when we learned them privately from a very venerable priest advanced in age, we were sworn and bound by oath never to betray them to anyone in writing, and we were shown the place in the writings of Calid the Philosopher where he attempted to describe them, but God forbade him. The desired effect of this hidden art is obtained from these two things and in no other way.

Therefore, my son, when you labor to follow our words, which are perfectly true, and in no way ambiguous, proceed along this royal road that we have shown you in such a way that you deserve to understand and look upon these two things by the inspiration of God, as if someone had told you in secret. For if your heart is truly righteous and open to Him, He will bring these two matters to you at the right timer ds He did me, an unworthy priest: I will pass them on to you privately, like a father leaving a bequest to his son. From the beginning of the work, therefore, remember those two things, which are only passed on by those to whom God the creator has granted a natural death; and mindful of this, proceed carefully, in great fear of God and reverence of nature, exalting your material step by step, so that your shoot may grow, and be brought from its white color to a red one, and your earth be duly dried.

Soon you will hasten to the end of late Spring and turn all the clouds that were in vapor form into rain so that they water your gold, lest the growing plants die from too much dryness. For four ordinary days after the rain maintain the same degree of heat on the watered earth in your egg in the bath on the dung, and note carefully how much vapor ascends again and of what kind it is.

On the fifth day, however, have ready clean sieved cinders in a new jar, about the same amount as there was dung in the bath, and place three or four folds of wet cloth around the joint that joins the dish to the oven of the athanor to loosen the adherent lute more easily, and when it has softened, take the whole warm bath out of the oven, and immediately put in your dish of sieved cinders. Join it carefully as before to the walls of the furnace with a suitable lute, and apply a heat like that of June, or the second grade, having the cinders well warmed, but not sufficiently hot as to burn your hand if you touch them.

While the glass is still warm from the bath and when the cinders are hot, place it on the hot cinders, so that there is three fingers depth of well compacted cinders beneath the bottom, and pack it roundabout with compacted cinders to the depth of its contents, or a little more; so early summer has its beginning, and the heat making the new shoots glad requires skillful management. Be careful that the heat of the cinders is the same day and night, so that they do not burn your hand if you touch them, as we have said, nor should the bottom of the dish ever become glowing hot from the fire.

This is the heat, of hottest June, of which Maria the prophetess and the Turba Philosophorum[24] speak, saying Digest with a slow fire that is like the heat of hottest June. And when you have regulated the heat like this for four to seven ordinary days, you should know that your crop is now at the stage where twigs come from shoots, and umbels and clusters from grass, plants and other kinds of vegetable.

The signs by which you will know that the umbels or twigs are being produced are these: The red that was previously blackish becomes a clearer red, appearing before the blackness is covered by redness, and very little vapor ascends from the earth. Remember, however, to observe the aforesaid rule that you should not convert your clouds into rain until you see that no more ascend from your earth, and that those that have already risen increase no further. Then you must continue the heat for no more than two days, then convert your clouds into rain that will water your field, so that it may bear fruit in due season.

---

[24] Volume 13 in The R.A.M.S. Library of Alchemy.

APPEARANCE OF FLOWERS AND FORMATION OF SEEDS

My son, let us give thanks to God the ruler of nature, for already our seed has produced a shoot, increased, grown, and brought forth umbels and clusters. Let us give Him thanks for at His nod our clouds have arisen and given timely rain that has made our field more fruitful. Let us give Him thanks, for the harvesting of the first seed is near, and the vintage approaches. For the field bristles with awns, stalks and ears as if armed with steel, and will soon yield the expected heaps of corn, and the vine has given clusters that will soon be turned into flowers and grapes.

O my son, give thanks to God who has deigned to bestow such wonderful secrets upon us. But I tell you, my son, that unless you follow the light of nature, and him who has shown you those two matters, you are close to error and straying. For what we have just called the conversion of clouds into rain, the ancient Philosophers called the tail of the dragon, and some said it meant multiplication and others the addition of fresh Mercury, not having studied the works of nature correctly.

Some have called our earth water, and when they told us to water our earth frequently, and even submerge it, they did not welt understand that this watering must be done by the natural exhalations of

283

the earth, and not by the addition of foreign waters; nor, although all the Philosophers cried it aloud, did they understand their words. Be careful not to submerge our earth, lest the seeds perish in the flood.

Therefore, my son, you should be careful, and follow the lead of nature in all things, and when you have seen a clearer red colour as the sure sign of the fourth stage when the ears emerge, as we said, apply the same heat, or a little more, for the next four to seven days and nights, so that attractive flowers swell out from the ears and umbels, and show to what species they belong. For the sure sign of those flowers will be a red mineral colour, showing a metallic brilliance, and the vapours that previously ascended in quantity will be much diminished.

Consider, my son, have you ever seen a flower of the field that did not bear witness to its genus and species? Is not the rose known by its flowers, and the eyebright by its (flowers)? Therefore, will not a metallic red signify the flower of a metal, and promise that a metal is to come? Of course.

This is the sign of early summer, producing both flowers from ears, and new ears from growing stems, and with the continued heat of July, the seed-forming operation, moving and setting in order, brings flowers to perfection.

Come then, my son, when you see the flowers of plants and vegetables, they are a sure sign of, fruits and seed to come, according to the course of nature (for we do not speak of monsters); raise (the heat) to that of a fine August, and after your metallic flowers you may be sure of seeing the seed of metals and minerals; but do not increase the heat much, only to that appropriate to the beginning of August.

When you have seen the flowers of your crop appear, apply the same heat with the cinders for three days and nights, and convert whatever vapors you see ascend into rain, but not so copiously as to destroy the flowers before the seeds have formed. On the fourth day have clean washed sieved sand ready, heat it on a gentle fire, and having taken the cinders out of the dish in the oven of your athanor, put this sand into it, not too hot, and into this place your glass vessel, not cold, but as hot as the sand, and surround it with sand to the depth of its contents, and have three fingers' depth of sand beneath it.

Having done this, heat the sand for one week, or until you see the red of your material shine golden yellow, with very little or no vapour rising.

When you see this, give thanks to God with hands raised to Heaven that late summer has given its fruit, and that your seeds are furnished with stems, the two signs of this being, as we have said, very little exhalation of vapor, and an almost golden yellow color in your material. But if some moist vapors still arise, do not be concerned. Do you not know that when fruit and seeds are formed, they are not to be plucked at once, but that one must await their ripening, which perfects them? Do not the figs formed in one year ripen in the next? Do not grapes, formed in June, or sometimes May, finally ripen in the month of September? Therefore restrain the hastiness of your desire, my son, and do break off the work before it is perfectly ripe. Be patient and magnanimous, and await the determined time.

COLLECTION OF THE SEED

Behold, my dearest son, your longed-for harvest approaches. Be glad and give immortal thanks to immortal God, who allowed us to understand these things, and directed my words to true and open speech on the matter. And you, my son, adore Him as a suppliant, so that when you begin the work, He may consent to show you those two hidden matters, so that you too, a student of this exalted philosophy, may be worthy to come to the precious banquets of this secret feast, and be able to pluck the golden apples of the Hesperides in the garden of Tantalus, having first, as the fables warn us, put the watchful dragon to sleep.

But since the garden is surrounded by a solid wall, how will you enter? How will you recognize the tree that bears the golden apples? How will you pluck them? O my dearest son, do you not see that the garden has been made on a high and barren mountain, whose bottom part is ever loud with the warring powers of discordant winds, where a kind of insatiable chaos nurtures the perpetual fight of heat and cold, and the middle part contains red and black dragons, that war continually with the insatiable chaos and the winds, and when the fury of the red dragons has been increased by the stimulation of contrary winds, they perish in their own heat, and chaos receives their corpse?

Nor does the eternal battle ever cease, for once the red dragons have died, black ones succeed them, that blaze up and take fire with their fury, and when they die like the red dragons, others at once follow, as we said. And this battle never finishes, for chaos and the winds never cease, nor do the black dragons ever lack issue.

The lord and guardian of the mountain, conquering the offspring of the black dragons, enters the middle of the mountain at certain times.

The mountain peak appears in three different ways during a single year, for in winter it blazes up like Mount Etna with hot sparks, in spring, the fire being first extinguished, it is flooded with stagnant hot waters and reedy marshes. In summer, when the hot waters have been completely absorbed, there are warm ashes here, which are sometimes so hot that you might believe they are being burned by winter's fire. In autumn the ashes in their turn disappear, and fine sand, very hot, sometimes to red heat, takes their place, which I take to be the remains, or at least the base, of those hot stagnant waters of spring.

It is in this mountain, my son, that I think you will see the royal gardens of the Hesperides, where there are golden and silver roses, and golden and silver oranges, bearing their fruit each year. However, the way to this garden is difficult, to enter more difficult still and to pluck the golden and silver apples by far the most difficult of all.

For the mountain's practice is to admit no one except in the cold of winter. Approach therefore in winter time, and do not shiver in the cold, for you will hardly be able to bear the heat of the fire that is at the entrance to the garden. On the summit of the mountain you will also see a high tower guarding the garden, with two ramparts that are all ablaze. You must be willing to enter the garden, especially to tame the bulls breathing fire through their nostrils, and pass through the fiery ramparts. It may be achieved with great labour and loss of time. It is certainly such a great task that you will not be able to pass before the end of winter. Moreover, danger from the flames of the huge fires and the heat threatens greatly.

At this entrance are to be acquired the medicines that Medea gave to Jason when he tried to enter the garden in former times. But if, dear son, you cannot find this medicine, work diligently to find a way to pass unharmed through the fortifications mentioned, for if you pass them by without going through them, you will never be able to enter the garden.

But I was seized by great lust and desire for the garden, and when I saw that I could not pass through the flames unscathed, I was unwilling to leave, expecting someone to come who would tell me how to pass, or that perhaps the heat of the fire would lessen. Now nobody came to me, and bye and bye winter ended at the entrance, and behold, suddenly there was a great commotion in the tower, and the flames began to diminish, and shortly afterwards they were extinguished, and the tower and the ramparts were partly swallowed up. When this occurred, at that very moment when the tower stood still, I ran towards the garden, somewhat astonished at what I had seen happening.

When I had got near the garden f was prevented from entering it by hot stagnant waters everywhere, for the garden was in the middle of the hot stagnant waters, surrounded by a hard transparent wall, and the garden and the waters themselves were surrounded by a brick wall. However, I saw three narrow steps in a footpath, whereby there was hope of access, so I began to follow it without delay, and the brick wall opened for me. When I was on the first step a certain evil-smelling and offensive rottenness forced me to delay.

The heat was also conducive to putrefaction, nevertheless, when I finally overcame it, it gave access to the second and third steps, upon each of which something occurred to hinder me; but when I began to think I was near the garden, the mountain was suddenly shaken by a tremor, the waters were swallowed up, and a deep ditch around the brick wall was left, the bottom of which was seething all over with hot ashes. Three lines of channels were around the garden in the middle of the ashes, of different heats, so that the one nearest the garden was much hotter than the others. I struggled, however, until

I reached the bank of the third channel, when the mountain was again shaken by a tremor, and covered in a dark layer of ashes, then hot sand appeared in their place, surrounding the garden on all sides in a single heap.

At this I began to have great hope that the end of my toil was near, and it, soon happened, for as I was looking from the heap at the beautiful flowers in the garden, I saw many wonderful things, of which it is not proper to speak, and I was astonished and fell into such a stupor that an old man who had entered the garden could scarcely rescue me from it. I also saw that he had seven keys in his hand, with which I thought he then opened the locks of the doors, while I stood amazed on the mound.

This venerable old man then led me to a tree of golden apples, near which a dragon that had been recently kilted lay dead, and the golden apples were stained with his blood. My soul burned with desire to pluck the golden apples. The old man, recognized this, looked at me calmly and said. Lay aside, my son, the enticements of earthly desires, for these fruits are given only to those of divine disposition. At his words I shook all over, for I had not heard such a voice before, and at it I became as it were changed, and it seemed to me that I understood many more things.

His appearance also seemed to me to change to
something else august and terrible, and I recognized
him as the lord of the garden, not the gardener.
Suddenly I was greatly afraid that my rashness in
thinking that I could ever furtively enter the
garden of so great a lord would be punished. But
while I was anxiously turning these things over in
my mind, held by fear, desire and hope, he put out
his hand and plucked some golden apples, and looked
at me and the apples by turn, and said, This is the
garden of good fortune and wisdom that we planted
for the delight of men; to keep out wild beasts we
surround it by a strong wall; and when we saw that
men sought it by fraud and guile, we allowed no
intelligent beings to enter, unless they were the
just, pure, honest and good whom we had attracted
ourself; and if we judge them to be constant we
bring them here after many labours, and send them
gently away, enriched by gifts of this kind.

So saying he gave me the apples that he had
plucked, which I, falling down before him, took up
with as much reverence and respect as I could, and
hid in my bosom, rejoicing. And while I was
preparing to render him the thanks that I felt in my
soul, he said, this is not the end, my son, but
follow, and he led me into a workshop where the
golden apples that required to be cleaned of the
blood of the dragon had been washed and purified; he

took a certain white sparkling powder from a wooden
box and gave me some, saying, This is the powder
that cleanses all stains, and revives the dead, go,
and hide it secretly, purify the besmirched (apples)
with fire and this powder, cultivate your earth, and
sow the purified (seed) and it will grow, and your
earth will yield you much fruit.

With these words he disappeared from my sight.
I was amazed and astonished and when I had come to
my senses I thought I had been roused from a dream,
and tired and weary as if from a long journey and
much labour, I fell into bed, and it would all have
seemed like a dream to me, except that I held the
golden apples and the powder in my hand, and
remembered all that had been said so exactly. But
whether I truly entered the garden, or whether I was
carried thither in a vision, or whether I saw it in
a dream, blessed be His name forever Who saw fit to
open to me the great mysteries of nature, and did
not withhold His gifts from me, an unworthy sinner.
Praised, blessed and exalted be Father Son and Holy
Spirit, one God for ever and ever, Amen.

CONCLUSION

You have, my son, the whole process of our
work, without any abridgement or any superfluity,
rendered in an accomplished style of narration.
Prepare your heart, therefore, that you may find

favour in the sight of God, for it is the gift of God, partaking of the mystery of the oneness of the Holy Trinity. O most famous science, that is the theater of universal nature, and its anatomy, earthly astrology, argument for the omnipotence of God, testimony to the resurrection of the dead, example of the remission of sins, infallible foreshadowing of the judgement to come, and mirror of eternal blessedness.

There is no undertaking more excellent than this, seeing that it contains and embraces all the sciences, and is itself contained by none of them. To your ineffable Majesty, be given everlasting thanks praise and benediction, for You have not hidden your favours from me, You have revealed to me the mysteries of this secret art, therefore, blessed by Your name for eternal ages. Amen.

The foolish man will not recognize it, nor the simpleton understand it, *Sugar to the Parrot, hay to the Cow.*

END

The True Method of Confecting the Stone of the Philosophers

A Letter to his Religious Superior by:
Monachus Efferarius
Extracted from Theatrum Chemicum vo. III, Page 143
Translated from Latin by Madeline Wright

There are two principles governing this art. The first thing to note is that this substance, which has been written about by all the philosophers of antiquity, is living silver of the wise. It is known by other names; gold, medicine, the philosophers' stone, the elixir, and countless other names, as will be clearly understood in the discussion that follows. Moreover, this physical medicine is called gold, as Geber explains: "The thing that converts and transforms natural metals into true gold is gold". And so he correctly says that "what makes gold, is gold; but philosophic medicine makes this, therefore, etc." And it should be noted that there are fundamentally two principia of this substance, namely matter and agent. As one of the philosophers explains, the matter is living silver and sulphur- or arsenic, which is the same thing. Take this well into account, reverend father, since - as the philosophers say - living silver is not the principium of this substance, either in its own nature or in the nature to which its own mineral has reduced it. It is the principium only in the nature of which it is brought by artifice. The same is true of sulphur and whatever sulphur is combined with. The reason is that no one has found anything in silver minerals or in any other minerals that can become living silver in its own nature, or that can become silver alone. Much less has anyone found any single thing separate from these, residing in its own mineral and its own nature. This living silver and sulphur, as I have already told you and explained to you, has never been dealt with anywhere in its entirety. It has been produced or deduced as a specific aqueous nature, extremely subtle, clear, bright and pleasing, known to the Philosophers as living silver. It has also been produced as a specific earthly nature, which is known as sulphur made by artifice. The philosophers kept it marvelously hidden, wanting it to remain secret. They wrote of it in their books in fragmentary passages and only rarely. And so, reverend father, I plan to have nothing to do with the subject of this artifice. Therefore you will not be able to do any experiments, nor would it be a good idea for you to have someone else do experiments. For one thing, it would take a great deal of space to tell you everything you would need for this particular artifice. It is time-consuming, expensive, extremely difficult, and very dangerous for those who do not have much experience.

It is important to note that the philosophers said that this living silver and this sulphur were one and the same substance. "In fact, it is such a substance that it is only one substance, issuing from one substance, as I have explained elsewhere." The proof that it is one substance is that they sometimes call living silver sulphur, and sulphur living silver, as well as both at the same time: living silver and sulphur. All the philosophers, in many specific passages, say that it constitutes one substance. Lucas says: "Have no doubt that the basis of this art, for which so many have laid down their lives, is one single thing." And Diomedes says: "We make use of a truly awesome nature, since a nature can only be added to in its own nature. Do not introduce anything foreign to it." And Bocosen says: "Be careful not to introduce anything foreign, for our art has no need of more than one substance, as Geber has noted above." And Astanus says: "There is only one nature than can overcome all things." Similarly, Pythagoras says: "Though the substance is known by many names, its true name remains one and the same." The philosophers have made many comments like these; it would take too much space to list them all. They are all designed to show and to prove

297

that there is only one substance on which nature can exercise its action in this operation. This substance, as I have explained above, is the living silver and sulphur that I have pointed out. But someone may ask why, if there is only one such substance, the philosophers have called it by so many names and compared it to so many other things. There are many reasons for this. Diomedes says: "They did this because they are so foolish and ignorant that they did not recognize the substance." Similarly, Morienus says: "These hateful men piled up names and misled later generations. They even call themselves hateful, since they hate for anyone but themselves to profit from this knowledge." Similarly, Pythagoras says: "They have given this substance various names, because of the excellence of its nature." Bonellus says: "They multiply the names, because in the operation of this substance appears every color that can be imagined. So they divide up its names according to the different colors. All the elements are contained in the work itself; therefore they gave the substance the name of each element in it." Since there are four elements in the substance, Orpholeus says: "According to my teaching, remember that you must mix together pure, crude, unmixed, and direct elements over the fire. Do not allow the fire to get too intense before the elements are conjoined and

bound to each other. Therefore elements that are
cooked over a slow fire are most prized and most
ready to be turned into other natures." That is
Orpholeus' comment. Also, the philosopher says:
"Convert the elements, and you will find what you
are looking for. To convert the elements, as another
writer says, is to make what is moist dry, what is
dry moist, and what is fleeting fixed. Then you have
the four elements that are in this single substance,
which are visibly manifest and can be extracted
naturally. Therefore the philosophers call it water,
the four bodies, and the four natures." Hermes says:
"This water is celestial in nature, causing the
separation of the elements that exist in bodily
objects near it. Then it compounds them once again
and reduces them to a unity." Nimidus says: "This
perfect compound is made of all four elements. But
the multiplication of names has been a cause of
error for men who work with unworthy materials --
for example, salts, alum, urine, dung, human blood,
sulphur, living silver occurring in nature, in
marchasitis, and in many other things. They do not
even know that nothing can be found in a substance
that is not already there, as Geber says. They do
not understand what the philosophers have said in
their writings using simile and metaphor. They think
that their art is beyond all price and so inviolably
secret that it must never be revealed to anyone. The

philosophers have said so many things about this water or this substance that it would take too long to relate them. Therefore you have the material principia of this substance, or the material cause, which is called the medicine of the philosophers: living silver and sulphur. What remains to be discovered is the agent cause. What acts on this matter or moves it to corruption is the heat that is the instrument moving it to putrefaction. There is no other agent in the world. This is what Alphidius the philosopher said: "My sons, understand that the agent substance in the entire world is this one thing -- heat. Without heat there is no movement or action. The very root of the disposition of matter is heat, though there are many degrees of heat or fire. One must determine which fire is present, and in which degree. The present fire is undoubtedly the fire of heat from horse dung. It should be cooked in the kind of fire which I will describe to you. It keeps itself concealed in moist horse dung, which is the fire of the wise, both moist and dark. It is warm in the first degree and moist in the second. All the philosophers spoke about this heat metaphorically and with many errors. They likened it to the heat of the sun and the natural heat of a healthy person." Mesig says: "The substance congeals in the heat of the sun. And so fools who use different kinds of fire are misled. They do not understand the words of the philosophers, since they do not know that the generation and procreation of natural things can only take place at a very moderate and even heat, never an excessive one."

Method of mixing

Now that we have seen the natural principia of this substance, we turn to the method of mixing or conceiving them. It must be noted that the material principia of this substance, which nature uses for its action in a marvelous operation, are living silver and sulphur, as has been pointed out. Each of these is of the strongest composition and most uniform substance. They are so closely joined in the smallest parts that no part can be loosed from another when they are resolved. On the contrary, whatever part runs for this resolved because of the unity of form which the parts have each to the other even in the smallest fragments. They are resolved by the even, agent heat in their own nature, according to the requirements of their essence. It is important to notice that the sulphur and living silver is converted to an earthly nature, and that from both the earthly natures a very fine smoke resolves out and is multiplied by the heat in the vessel. And this double smoke is the matter of metals or medicines, or the components of the philosophers' stone. It is converted into the nature

301

of this same earth by smoke and moderate heat in the vessel of its decoction. Then it takes on a certain fixity, which makes it possible for the water flowing through the vessel to destroy its porousness. And so it becomes viscous throughout, and all the elements, each in its own natural proportion, arrive at a unity. They mix together even in the smallest particles until they achieve a uniform mixture. This is done through successive decoction, every day, at the most moderate heat in a vessel, until the particles thicken and harden. Then they are medicines, or metals, or the philosophers' stone. So Morienus says: "The disposition of the wise and the transformation of natures is the remarkable mixture of these natures by means of the hot, cold, moist, dry, and subtle dispositions." That is the only argument of the wise. Note the entry, submersion, fixity, connection, thickening, conjunction, joining, composition, and mixture, since when these are combined in order all other things are mixed together. According to Democritus, all these things are one and the same. There is a mixture of things that can be mixed, that is, of elements: for these are the first principia of every single thing that is mixed. We cannot know either the manifest or the hidden nature of anything that is mixed unless we know how to mix or compound those elements. Hermes says: "Children of the wise, understand the science of the four elements, which follows its own reasoning in a sort of hidden revelation. The hidden revelation of the elements means nothing unless it is compounded, which cannot come about as long as it is passing through its various colors." Note that this takes place in the smallest--that is, indivisible--particles. For the smallest particle is the one that is indivisible. If it can be further divided, it cannot be the smallest. So it is clear that the mixture of elements takes place in the smallest particles of a body—namely, the indivisible particles. For an element consists of the smallest, simple particles of a body.

Concerning the effects of the principia

Next I shall set forth the effects of these principia, namely living silver and sulphur. As proof of them, one should take into account that there are several degrees in the operation of this substance. Lucas says in the Turba: "It does not need several substances, but only one, and that one substance can be changed into another nature in each single degree of any work. The kinds of degrees correspond to the various proportions of elements that can be mixed and that come forth in the operation. "The philosophers named each operation in each degree according to the order it attains in nature and in its process of generation. They called the first degree of operation iron or Mars; the second copper or Venus; the third, lead or Saturn; the fourth, tin or Jupiter; the fifth, silver or the Moon; and the sixth, the Sun or gold. Metaphorically, they gave countless other names to these metals, all in order to keep the knowledge secret. These, in so far as they are generated from the same first principia--that is, from a single first matter, which is living silver and sulphur – are rightly named the effects of those principia, living silver and sulphur. The philosophers have defined each of them separately and specifically according to their various properties. Each is dealt with separately because each has a different composition and generation in the process of creation or production. One should note that the philosophers sometimes give the name of iron or lead to the substance they know as their specific gold, and so on with the others. Also, conversely, they call gold iron or lead, and so on with the others. And this is true in various respects, since the resolution and corruption of one thing is the generation or cause of another. To that extent, the effect resides in its cause. So they were able to say that gold is lead, and so on with the rest. You can see how men who work with natural metals, calcining, dissolving, and congealing them, made their mistake. They believed that they were making philosophic medicine from these metals, because the philosophers say that gold can be brought out of iron, lead, and the rest. Also because the substance they call gold or the medicine of gold is produced from the resolution or corruption of other metals. For all the metals mentioned above are generated and corrupted in this very operation, and the specific gold of each is generated from them. So they were able to say that iron and the rest are gold, since the cause is in a way its own effect. For these reasons the other metals can exchange their names back and forth, in so far as the corruption of one is the generation of another. It should be noted that the philosophers call all these metals living silver and sulphur, because they all come into being from those two. And the philosophers called these processes of generation of metals complexions. They are all unbalanced compared with gold. They said that only the complexion or perfection of gold is balanced. And they looked for only one complexion or generation among all of them, namely the balanced one, the one belonging to gold. Johannes wrote about it: "The elemental mixture is only this one, for its body remains intact even when it is modified." Note that he speaks about moderation of the four natures--heat, cold, moisture, and dryness. When no one of these natures is greater than any other, the body is rightly called balanced. What is true of one is true of the others. And it is also rightly called intact, that is, sound, and pure of every cause of corruption. Note that tin and lead meet these standards. It is itself gold or the medicine of gold. Geber says: "Tin is the purest lead. Its two components, living silver and sulphur, have a quality of fixity and thickness. It is not a matter of quantity, since it surpasses living silver when it is mixed." Hermes and the Sons of the Philosophers say: "There are seven bodies whose first mode of generation is not gold, but the intention of nature, since nature always seeks and moves toward what is best, noblest, and most perfect. It is first in nobility and worth. Its chief and king is gold, which earth cannot corrupt nor fire burn nor water alter. That is because its complexion is tempered and its nature is stable in heat, cold, moisture, and dryness. There is no excess or defect in it. And so the philosophers exalted and magnified it, saying: "Gold has the same place among bodies or metals that the Sun has among the stars." We must still determine the principia of artifice of the substance we are seeking. Note that they are called principia of artifice even though they occur in nature, because nature can only operate through their aid and artifice. They are the methods of operation. Only by their agency can the substance we are seeking be generated and brought to actual existence. The first method is sublimation, the second descension, the third distillation, the four the calcination, the fifth solution, the sixth congelation, the seventh fixation, and the eighth iteration. And there are countless others like these. These methods seem different, but they are all really the same. Once the philosophers were looking at matter which was in a vessel, caught the heat of the sun and exhaled or evaporated in the form of a very fine smoke, it rose to the top of the vessel. They called this ascension or sublimation. Afterward, they saw the material which had risen descend to the bottom of the vessel and called it distillation or descension. When they saw the matter thicken, turn black, and give off a fetid smell they called it putrefaction. Much later, they saw the black or dark color fade, and a paleness like that of ash take over. This they called incineration or whitening. Morienus says: "The entire teaching is nothing but extraction of water from earth and pouring water back to earth until it putrefies. And when the earth is putrefied from the water, and later everything is purified with the help of the instigator, the entire teaching is complete. When they saw the earth mixed with water, and then saw the water gradually decrease and the earth increase as slow decoction proceeds, they all pronounced that ceration had taken place." The philosopher says: "The earth is cerated with water, drinks it in, and then dries in slow decoction by the heat of the sun. Then the entire matter turns to earth." Again, he says: "It is complete if it has been changed into earth." They saw that the entire matter arrived at a kind of dissipation, and that it somehow was reduced to a solid substance. Since it no longer flowed but stayed still and firm, they said that it was in perfect congelation. Plato says: "Dissolve the stone, and then congeal it with great care, as you have been instructed. Then you have almost the entire teaching." In another passage, he says: "Take the stone, put it in a vessel, and roast it over a slow fire until it breaks apart. Then cook it in the heat of the sun until it congeals. Then you will know that the entire teaching is nothing but how to make a solution of substances and a perfect and natural congelation." He also says: "Dissolve, congeal, and then you will understand the entire teaching." The philosophers saw the matter perfectly congealed and thickened, so that it could never again turn to liquid or smoke by any means. They said that it was truly fixed, because they saw that the same congelation and thickening, or fusion, had come to perfect desiccation and whitening by a long decoction over heat. This whiteness was purer than any other; and they called it perfect calcination.

When they saw that the matter had come to a stable heat and had turned countless different colors, which can only happen in resolution of the matter, they called it solution. In such resolution, elements cease to be continuous. They are both active and passive alternately. So the philosophers call these elements mates. And the fools who believe that physical medicine can be created from any substance are grievously wrong. The philosophers say that the alchemical sons who put their trust in all their dissolutions, sublimations, conjunctions, separations, congelations, preparations, grindings (contritions), and other such processes are obstinate in their errors. And let us hear no more from those who preach that there is any use or place in this operation for any other gold than ours; any other water than ours (which is also called extremely sour vinegar); any other dissolution and congelation than ours, which is made over a slow fire; any other putrefaction than ours; any other volatile substance than ours, whether a spirit or anything besides our own living silver and sulphur; any other alum or salt than ours, which in its whiteness is called the Flower of white salt. And let us hear no more about any other egg or any other human blood than ours, or anything else extracted from any vegetable matter, or human being, or brute beast. It is always possible to make an error in that regard, since our stone has as many names as it has substances. Let everyone hear what so many philosophers have said: nothing but a human being can be produced by a human being. And the same is true of animal and vegetable creatures. Many have made false application of their knowledge, to the point that they themselves are deservedly known as fakers. And let all those who believe that our work can be accomplished with the dust of the brute basilisk be silent. Perhaps they have been misled by philosophers who said that our stone smells like the air coming from a tomb. So perhaps that is why they chose the basilisk, which reportedly has a fetid stench. If, on the other hand, you find that our substance is being properly nourished, like an embryo in the womb (as some of the philosophers say), this is because it has been decocted over moderate heat. The length of time is not the reason. The substance congeals at a moderate temperature; when someone claims that it takes its nourishment from eggs, you will see, etc. Note that the entire teaching and intention of the philosophers is simply to divide, purify, and unite. Again, entire perfection is nothing other than perfect solution and congelation. Note that the entire procedure consists of fire and heat. Entire perfection is simply this -- to convert the elements. The entire procedure is simply to cook, roast, fit onto thin tablets, file, cut with scissors, trim, putrefy, incinerate, water, separate, divide, purify, whiten, redden, dissolve, shred, wax, mix, heat, pluck, sift, irrigate, moisten, weaken, absorb, starve, dehydrate, boil, mince, pour, cut with a fiery sword, pound with a hammer, separate the soul from the body, pour through, convert body to spirit and spirit to body, mate, impregnate, sublimate, fix, descend, calcine, dissolve, corrupt, and coagulate. This is all. There is nothing more, save to distill it in fiery heat in a cucurbit and alembic. All this is kept as a great secret of the art, as the following verses show:

The five substances are one, in only one vessel, and only one boiling:

There is only this one recipe for this one substance.

This is the philosophers' general rule: that a moist substance can thicken only if there has first been an exhalation of very fine particles from it. Also, its thicker particles must first have become finer. This takes place when the moist is stronger than the dry in composition or mixture, and there is a genuine mixture of dry and moist, so that the moist is tempered by the dry, and the dry by the moist. This can only be done by mixing together the viscous, moist part and the fine, earthy part in their smallest particles until the moist and dry are completely homogenized. The resolution or exhalation of the fine, vaporous particles or fumes cannot be done all at once, but only gradually and over a long period of time. The cause or reason for this is that the substance of the principia is uniform. The resolution of excess moisture from them must not be done too fast, since the moist and the dry are scarcely separate. The mixture between them is strong. There is one continual, even cause of the thickening or composition of the metals, because of the strong union of the viscous, moist part and the fine, vaporous part.

The End

The Treasury of Philosophy

By Monachus Efferarius
Theatrum Chemicum, Vol. III, Page 151
Translated by Madeline Wright

There is one who created all things out of
nothing, and to whom all things belong -- sky,
earth, and sea. He bound together all harmony and
all dissonance, and of his great goodness willed to
cure disease with his medicines. Wise men in ancient
times fully understood two methods and wrote them
down in their books. One method is true; the other
is deceitful. The true method they set forth in
obscure terms, so that only their disciples could
fully understand it. They concealed it to keep
impious men from seizing knowledge and using it for
profane ends. They would have to pay the penalty for
their sins. So, dear reader, do not share any of
this knowledge with an unworthy person. Keep it
secret, like a true philosopher. When you finally
test it out in a real experiment, you will love and
esteem it all the more.

The false method they set forth in very clear
terms. I am not going to repeat the method or the
errors and what causes them. But listen and
understand, dear reader, and may God illumine your
mind. Know that our science is the science of the

four elements, the four times, the four qualities, and all of these in reverse. All philosophers agree about that. And know that the four elements exist in everything under heaven. You will know this by their effects, not by observation. So the philosophers have handed down this science in the guise of a science of the elements. And they have caused it to operate without understanding the literal account of the operations. They have tried to do operations with things like blood, hair, and eggs. I have tried too, and have been reduced to a daze. I nearly despaired of science altogether and discarded all its teaching. I blamed others for the feebleness of my own intellect. At last I came to my senses and began to ponder what Avicenna had pondered: "If something exists, how does it exist? And if it does not exist, how does it not exist?" Therefore I discovered that the matter and seed of every metal is Mercury that has been boiled and thickened in the bowels of the earth. It has been boiled at sulphuric temperatures. According to the different varieties of sulphur, different metals are produced in the earth. Their first matter is one and the same. They differ only in their inessential activity and whether they have been boiled in a greater or lesser fashion, at moderate or excessive temperatures. Some have been burned, others have not; and all the philosophers are in agreement about this. It is certain that everything comes out of or from whatever it is dissolved into. For example, ice becomes water when heat is applied. Clearly, then, it began as water. In the same way, all metals become Mercury, since they were made of it in the first place. Later I shall show how to convert them into Mercury. Once this is understood, the saying of Aristotle in Book IV of the Metaphysics makes sense: "Alchemists must understand that specific types of metals are immutable." That is true, as long as they are not reduced to their first matter. Such a reduction is possible, even easy, since everything that lives and grows can be multiplied. All plants and trees and animals demonstrate this. Out of one grain come a thousand grains: out of one tree come a thousand little branches, and out of one man comes the entire human race. Just as everything is increased in its own kind, so a metal can increase its substance by its own agency. And it makes no difference, as Aristotle says, whether this is done by natural or artificial means. All metals live and grow in the earth. Therefore it is possible to increase and multiply them to infinity. But it can only be done through a more perfect agency — the consummate medicine of the perfect generation of metals, which is called the elixir of the philosophers. The only way to acquire this elixir is through its own medium. Notice that it is the nature of this medium to be stable and free of extremes. The extremes are sulphur and mercury, and the finished elixir. The better and more perfect ones-- also the more accessible--are the ones that have been more thoroughly purified, decocted, and digested. Dear reader, do not err: whatsoever a man soweth, that also shall he reap. It is obvious what the stone becomes, and what kind of medicine it makes, if nothing extraneous is added and any excess is removed. Only what is by nature close to a substance is suitable to it. Now, dear reader, I shall explain the sayings of the philosophers, the obscure words of the wise that are hidden in parables. Then you will know that I understand their words and faithfully represent them. First comes the process which the philosophers called solution. It is the foundation of the art, as Maria says: "If you join gum with more gum in a true matrimonial union, you will have a torrent of water." The philosopher Rosinus says: "Unless you can turn bodily substance into incorporeality, you labor in vain." Parmenides also dealt with solution in the Turba: "Some people have heard of solution and believe it can be done without the body it is joined to. But they are permanently one. This is not the philosophers' solution of a watery cloud, but solution or conversion into the water from which it was made in the first place -- that is, into Mercury. In the same way, ice turns into the liquid water it once was. By the grace of God, see now this one element called water, and the reduction of a body to liquid water. The next step is that earth is made out of water in a slow decoction, repeated until it is predominantly black." Avicenna says in his treatise on Humors: "Heat working in a moist body first produces blackness, as you can see in lime made by the common people." Menabdes says: "I bid posterity to make bodies incorporeal by dissolution, and turn incorporeal substance into bodies by gentle decoction. But you must be very careful not to let the spirit turn into smoke and vanish because the fire is too hot." Maria says: "Watch over it, and be careful not to let anything escape in smoke. Do not let the fire get any hotter than a July sun. Then the water will thicken in a gentle and slow decoction, and the earth will turn black. Now you have another element, earth. And the third step is to purify this earth." Morienus says: "This earth putrefies with its water and is then purified. When it is purified, the entire teaching will be directed with God's help." Hermes also says: "Azoth and fire cleanse the lato[25] and take away its blackness." The philosopher says: "Whiten the lato and restore its offspring lest your hearts should break." This composition belongs to all the wise, and is also one

---

[25] "Lato" is the impure body. -pnw

third of the entire work. Then, as it says in the Turba, join the dry to the moist - that is, the black earth to its water — and boil it until it turns white. The fourth step is to cause the water, now thickened and coagulated with earth, to rise by sublimation. Then you will have earth, water, and air. And this is what the philosopher says: "Whiten it, and sublimate it quickly over the fire until a spirit comes out of it. This spirit that you will find in it is called the bird or the ash of Hermes." Morienus says: "Do not hold the ash cheap, for it is the very diadem of your heart, the ash of the things that last." The book called Turba reads: "Speed up the procedure of burning, and after the whitening stage will come cineration or formation of ash. This is known as calcined earth, and it has a fiery nature. You now have the four elements in the proportions predicted: dissolved water, whitened earth, sublimated air, and calcined fire." Aristotle speaks of these four elements in his letter to Alexander on the rule of princes: "When you have formed water from air, air from fire, and fire from earth, you will possess the art of philosophy in all its fullness." And this is the end of the first composition, as Morienus says. Now let us move on to the second composition, which has to thicken, tinct, and bring the first composition to life. Calindus the philosopher says: "No one has ever been or will ever be able to tinct foliated earth with anything other than gold." Hermes has the same precept: "Sow your gold in white, foliated earth that has become fiery, fine, and airy through calcination. We sow it with gold whenever we put the tincture of gold in it. But gold can never tinct anything but itself to perfection. In actual fact, it can only be done by means of art." Raymundus says: "This stone of ours already contains a tincture in itself, naturally. It has been perfectly remade in to the body of magnesia. But it can be made perfect only by art and operation." Geber says in his Operation of Roots: "The purpose of the operation is to make the tincture of gold that is in gold even better than it is in nature. Also to make an elixir as in the allegorical fable of the wise, a clear condiment made of different species. It is the antidote and medicines to cure, purge, and transform all bodies into true silver and gold-producing substances." Listen to what Hermes says about whether we need gold alone and no other body: "The father is hot and dry in order to produce tincture. The mother is cold and moist in order to nurture offspring. So both gold and silver in themselves are very difficult to unite. Even when they join in a way that causes solidification into gold, they pour off very quickly." Maria says: "Take a projecting body or a clearly defined mound which has not putrefied and pound it with a stone until it takes on the tincture of spirits. Then set it near the fire. It will all melt quickly if you have cast over it its wife, silver. If there were anything else in our stone, the medicine could not flow out so freely, nor could it give a tincture. And even if it could, it could only tinct according to what it was itself, and the rest, including Mercury, would fly off in smoke. There would be no receptacle in it to receive a tincture. But our final secret is how to have a medicine that flows before the Mercury escapes. The conjunction of these two is absolutely necessary in our work." Geber says: "In a perfect teaching, the supremely perfect metal is gold. With its tincture of redness it transforms every other body. It is a yeast that converts the entire lump of dough into its own nature. Further, it is the soul that unites body and spirit. Just as the human body is dead and immobile without the soul, so any other body is unclean, earthbound, and lifeless without the yeast that is its soul. For the yeast of the prepared body converts the entire lump of dough into its own nature. And there is no yeast other than gold and silver that has been taken from the planets. Just as the Sun and the Moon reign over all the other planets, so these two bodies reign over all other mineral bodies. These other bodies are fittingly converted into the nature of the first two. It is proper to call it yeast, since without it the grains cannot undergo change." Raymundus says: "You will not be able to work these changes unless you refine it by art and operation beforehand." Hermes says: "Son, extract the shadow from the ray of light." So we must complete the preparation and refining of the yeast. It is like a fountain, for it arises perfectly with respect to creation, but not to operation. It must first be fed a little milk, then more, and finally a large amount. This is just what happens with our stone. Then take a fourth part of it--that is, one part of the yeast, and three parts of the unperfected body. Dissolve the yeast in an equal amount of mercurial water. Cook it all together over a very slow fire. Then coagulate the yeast until it becomes like the unperfected body. Let it sit in the orifice of a closed vessel, by the method and steps described above. Hermes has the same precept: "Dear reader, mix the equal parts together at the beginning of a fresh operation. Pound it until it joins as in a betrothal. Feed it until conception takes place at the bottom of the vessel and generation takes place in the air." Also, Morienus says: "First cause red smoke to capture white smoke in a strong vessel. This should happen by a firm conjunction without any exhalation of spirits. This is the fifth step. The sixth step is to join a fourth part of refined yeast with three parts of whitened earth. Absorb its water into it until the two bodies become one, and there is no difference in color." Morienus says: "Once the white body is calcined, put in a fourth part of the yeast of its gold. For gold is like yeast in bread, which converts the entire lump of dough into its own nature. Cook it in its water until it becomes one substance and one dry body." Maria says: "Let the air strike and congeal it and it will be one body. That is the secret of Scalia. Then the yeast can be introduced into the body, since it is its soul." This is what Morienus says: "Unless you cleanse the unclean body, make it white again, and send its soul back into it, you have not carried out any of this teaching. This is how yeast mixes with a changed body, not with an unclean body." Gasius, in his perfect teaching, says: "Stones cannot receive each other unless they have first been cleansed. A body cannot receive a spirit, nor a spirit receive a body - so that the spiritual becomes corporeal and the corporeal spiritual -- unless they have first been completely purified of all filth. Once they are cleansed, body and spirit soon embrace each other. Then a single perfect operation arises from them, because nature has remade them, and whatever was thick and gross is now refined." Astanus says in the Turba: "Spirits cannot unite with bodies until they have been completely stripped of all impurities. In

the hour of their conjunction the greatest of
miracles appears: the imperfect body takes on a
stable color by means of the yeast, since the yeast
is its soul. And the spirit, by means of the soul,
unites with the body, takes on the color of the
yeast in that instant, and becomes one with them."
This is the elixir, as Avicenna explained to the
philosopher Assis. It is tincted with its own
tincture, submerged in its own oil, and fixed in its
own lime. We discover its water as living silver in
minerals, its oil as sulphur or arsenic in minerals,
and its lime as lime in minerals. The work itself is
noble, abundant, and sublime. Whiteness is the mark
of the three in which there is no fire; and the
color yellow spins on four wheels. Maria speaks of
these wheels: "In that school there is nothing but
marvels. Four stones enter, and their procedure is
genuine. Therefore anyone who has a subtle mind or
intuition knows that philosophers tell the truth in
obscure words. For they say: 'Our stone is made of
the four elements,' and they have compared it to the
elements. It has first been shown how the four
elements are present." As Rasis says: "Everything
that the great creator has placed beneath the moon's
sphere has a share in the four elements. This can be
seen by the effects, not by observation. For the
stone is one single thing -- only one substance, one
root, and one nature." Hermes says: "In the Lord's

name, begin to understand the nature of the stone. It comes from the root of its matter, since it is both in it and of it. Nothing can enter into it that did not arise from it. In reality, it is not fitting to a thing unless it is naturally quite close to it. For every single thing loves what is similar to it." Therefore Plato says: "It is one substance and one essence: in it alone hot, cold, moist, and dry are present. And it has been called a minor universe, since from it, through it, in it, and with it exist all metals. Further, it is like a tree whose branches, blossom, and fruit are both from it and in it, and it is alike throughout. Every single thing can only produce what is similar to itself and of its own kind. And so this thing is one and the same, and whatever comes from it is one and the same, in no way different. The philosophers call this stone by the name of the body of each thing and each species." Pythagoras says: "It is called by every name, though only one name is proper to it." Hence the verse: "This one single Moon is known by every name."

And Phieras says: "Have nothing to do with the proliferation of gloomy names. Nature is one, and is above all things. All the different natures together cannot change that single thing. On the contrary, there is only one nature, and it causes itself to germinate." Therefore, as Diomedes says: "The nature we make use of is venerable indeed, since nature can be changed only by its own nature. Do not think you can bring in so much as dust or any other thing that is alien to it. All the diverse things together cannot change it, for it causes itself to germinate."

Maria says: "The whiteness and moist lime which come out are the one single thing from one single thing, and are the very roots of this art." The philosophers gave all possible names to it, yet it is only one thing, as Morienus says. In all truth I tell you that the multitude of names is the only thing that has led modern men into error. But let every wise man know that these names are nothing but colors appearing in conjunction. Consequently, you will not wander along the road of the work, even though philosophers have come up with so many sayings and so many names. Still, they refer to one thing only, one means of operation, and one change of temperature (or of color). And note that no such variety of colors appears in our stone or in the conjunction of soul and body, as Morienus has it. Just one change of temperature restores all the various colors. The philosophers have said that the stone is composed of body, soul, and spirit, and they are right. For they called an imperfect body a body, yeast a soul, and water spirit; and they have done well. For an imperfect body in and of itself is heavy, weak, and actually dead. Water is the spirit that purges, refines, and whitens a body. Yeast is the soul that gives life to an imperfect body. It had no life before; now it is brought to a superior form. The body is Venus, a woman; the spirit is Mercury, a man; and the soul is both Sun and Moon. The body must be melted into its first matter, which is Mercury, as Morienus says. Mercury can only be procured from the liquefaction of liquefied bodies. It is not common liquefaction, but the kind that lasts until you join them in true matrimonial union, when they are united and attain to whiteness. Note that the body is completely liquefied when blackness appears in decoction. Bonellus says: "When you see that blackness is about to overtake that water, know that the body is now liquefied. Then cook it in water over a slow fire until it dries up with its attendant vapor. It then becomes a thing which can be introduced into its own body. The spirit converts the refined body into itself and penetrates it. Therefore it is called aqua permanens and aqua vitae." Mundus says in the Turba: "Mercury is aqua permanens, and without it nothing can be made. Its strength is of the spirit and its blood is ground down. When it turns into spirit along with the body and both are mixed together, they are reduced to oneness. This occurs in the measure of strength in which the body incorporates spirit. The spirit turns the body into tincted spirit, as if it were blood. For whatever has a spirit has blood as well. Blood is the matural[26] humor that strengthens nature. And know that the longer it is boiled and purified in its humor, the clearer and better it looks." But, as Morienus says: "Nothing but azoth can remove the shadow from lato. It should be boiled in azoth until it turns it several colors and then as white as fish eyes. Then the good part comes forth and joins with the yeast. Note that the yeast is the fixed soul of the stone, and that it tincts, brings to life, and enfolds." Maria says: "The fixed body is made of material of Saturn. It encompasses the digestion of tinctures and fills it with wisdom. Without it this teaching will never come to full effect until the Sun and Moon are united in one body. For the entire artifice of this art, as Euclides says, is in the Sun and Mercury." When these are joined as one, they possess an infinite tincture. In a work it waits for a color that is redder than blood. When a little bit of this color is poured into white, it turns even a large expanse of white into yellow. You can test this by tossing blood into milk or water. Therefore, as Josephus says: "Mix fire and water, and there will be four: then make them all one, and you have reached what you were looking for. The body will no longer be a body, weak over a fire that is not weak;

---

[26] Natural maturity or being mature in a natural way. -pnw

and peace will rest upon it." From start to finish the preparation of these things consists of fixed water. It is honorable because it shows tincture in projection. It is also the intermediary between opposites, and is itself beginning, middle, and end. Whoever understands it grasps wisdom. Some of the philosophers have said: "Unless you turn bodies into what is incorporeal, and incorporeal substance into bodies, you have not yet found the rule of truth." And they are right. For the body first becomes water, and so corporeal substance becomes incorporeal -- that is, spirit. Therefore Hermes says: "Convert natures, and you will find what you have been looking for." And that is true. For in our teaching we first bring a delicate substance out of a thick one—water from a body. Afterward we bring dry from moist--earth from water. And so we really do convert natures, because we bring spiritual from corporeal and corporeal from spiritual." This is what Senior says: "There is a conversion of bodies from one state to another, from one substance to another, from infinity to potency, from thickness to slenderness, from body to spirit. In the same way, a man's seed is converted in a woman's womb from one substance to another by a natural process of conversion. At last a perfect human being is formed. From this process come the root and principium of a person, and no change can be brought out of the root by any process of division." As Aristotle says: "All generation comes from appropriate sources in nature." That is true, especially in the case of generation of metals. The philosophers say: "Let no foreign substance enter into it, whether dust, water, or anything else. If something foreign does enter, it will corrupt and destroy it." A king named Arabs has said: "Water can only adhere to something similar to its own sulphur, because it comes from it. So we make it into something higher, as we do with what is lower. Namely, the spirit becomes body and the body spirit. This happens in sublimation just as it does at the beginning of our operation. What is lower is like what is higher: it all turns into earth." Therefore Hermes says: "What ends up higher after sublimation becomes lower after descension, and what is lower after constipation is like what is higher after ascension. They all work miracles in this one respect: water and earth have a lower place." Air and fire rise higher. Water and earth conceive and nurture. Air and fire are active and conjoining. And all four come together in our stone, as Senior says: "The four elements are found purified in our stone. In it, water is fixed, air is stable, earth is at rest, and fire surrounds everything." They come together in the stone in spite of all resistance. And their four natures are generated in the stone, from it, and through it.

This is clear from the above premises, since our stone is made of all four elements. So the philosophers have said: "Our stone contains body, spirit, and soul. All three come from one nature, one substance, one water, and one root." They are definitely right. For our entire teaching is performed with our water. From it and out of it come all necessary things. It dissolves bodies, not by the common process of solution that the ignorant use to turn clouds to water. It is done by true philosophic solution, which converts clouds to the water they came from in the beginning. Socrates says: "The secret and the life force of every single thing is water. It dissolves body into spirit and restores the dead to life. It is the extremely sour vinegar that overhangs and overpowers all things. So pound our stone with this sour vinegar, and boil it in the same vinegar until it thickens. Boil it very quickly so that the vinegar will not turn to fumes and disappear. Then this same water calcines the bodies and reduces them to earth. It then turns these bodies black, white, and red in turn. It then transforms them into ashes, pulverizes them, and enters them." On this subject, Marchos the king says: "Our water dissolves bodies, then congeals them and turns them black. It also cleanses every body, takes away all blackness, and tincts whatever is black. It makes them white, and then tincts them to make them red. And it reanimates and brings to eternal life whatever has died. Therefore this water is praised, exalted, and proclaimed mistress of all things. Nothing else can perform its operations." Morienus also says: "Azoth and fire cleanse and purify lato. They completely remove any trace of darkness from it. Lato is an impure body, but azoth is mercury. And this water unites diverse bodies that have been prepared for the conjunction according to the method described. For fire cannot separate them. The water effects a marriage between body and yeast. It then turns their oneness into something else, and keeps them from being burned by fire. For the calcined and whitened earth seeks a higher place. It has become airy and spiritual. And whatever is airy and spiritual is incorruptible and able to penetrate." Hermes says: "Aqua aeris exists between earth and sky, and is the life of every single thing. It is the intermediary between fire and water through its heat and moisture. From those the water takes in air, for air itself takes in fire. It is like fire because of its heat, and close to water because of its moisture. Therefore it causes a matrimonial union of man and woman. In fact, every spirit consists of the fineness of smoky air. For every living, vegetable substance draws spirit and life from air. Therefore fire brings dead aqua aeris to life, causes a matrimonial union, and

keeps the composite from being burned by fire." And
so the philosophers have said: "Convert water to air
so that life can exist. Life exists with life,
because it is itself life and the spirit that it
enters into." Therefore our water sublimates bodies,
though not by a common process of sublimation –
which is what idiots intend to do when they consider
sublimation better than ascension. So they take
calcined bodies and mix them with sublimated
spirits, such as sulphur and mercury with sal
ammoniac. Over a strong fire they bring about
sublimation. The bodies rise along with the spirits,
and then they say that both spirits and bodies are
sublimated and completely purified of any excess.
But they are deluded. Afterward they find them even
more impure than they were before. For art is weaker
than nature, as Albertus says in his book about
minerals: "The two extraneous humors have been
purged of the substance of sulphur both by artifice
and the genius of nature. Art cannot purge or
cleanse them as well. For the artifice of nature is
more sure and more sublime than any kind of art." So
our sublimation is not superior to ascension. The
sublimation of the philosophers comes from a simple,
lowly, and corrupt source. But it can become great,
exalted, and pure. So we say: "This has been
sublimated to Episcopus (bishop), in fact promoted
to Episcopus." That is to say, promoted to a rank of

dignity. So we say that bodies have been sublimated or refined, as well as transferred to another nature. So sublimation is the same as refining, and our water achieves all of this. Morienus says: "Water taken from a dead body which the soul has left removes any stench. And once it has whitened and sublimated the soul and guarded the body, it removes any darkness or bad odor from it." Albides also says: "Take things from their minerals, sublimate them to a higher place, send them down from their mountain peaks, and reduce them to their roots." Therefore, to sublimate is to refine something gross. Hermes says: "Sublimate the fine from the thick very gently and skillfully. Earth rises to the sky and then falls back to earth again. It receives power to penetrate what is high and fine as well as to remain in the heaviness of what is lower. This is how you should understand sublimation of the philosophers. There are many who have been wrong about it. So our water brings bodies to life and then deprives them of life. It leads them to their fall and then back to their rising. In the process of mortification it causes black colors to appear. They are then turned into earth by putrefaction. Afterward, many different colors appear before whitening, but they all cease when a stable whiteness takes over. It is like a grain of wheat falling to earth, which unless it dies, it

333

remains alone. So the seeds of everything that the earth produces change and putrefy, until at last corruption overtakes them. Then they germinate and increase like the very earth they take root in. So our water is nourished, putrefied, and corrupted. Then it germinates, rises again, and bestows new life on itself." Therefore Calidus says: "When I saw the water congeal itself, I was certain that this thing is true as claimed. Cook it along with its body until its moisture dries up from the fire and it is entirely dry. Then it collects its spirits and makes its dwelling in the root of its element. This will happen once you have mortified it and boiled the white body. It will then become <u>aqua spiritualis</u>, able to convert natures into other natures. It will also give life to dead bodies and make them germinate." Our water is the mother of many marvelous colors, since through it a diversity of color appears. This will take place especially in the sprinkling of water from a prepared and fermented body. An infinite number of colors will appear - as many as can be imagined. For the spirit is united with body and soul. The spirit is the dwelling place of the soul, and the soul is taken out of bodies by the tincture of water. Senior says: "This water is a tincture dissolved on a body, just as the tincture of tinctures is carried on a piece of cloth. Then the water recedes in desiccation, and the tincture remains in impression. It is the same way with water of the soul that carries the tincture. One can bring it back to its parched, white earth in foliated pieces. Hermes calls this water the gold of thorns, flowers, and saffron, since it tincts their calcined earth." He also said: "Sow gold in foliated earth. Then the aqua spiritualis recedes, and the soul, which is the tincture of the sun, remains in the body. It is like a fine smoke, imperceptible, appearing only in its effects. But its action is the manifestation of colors, and fire generated from fire and nourished by it. In fact, the soul is the daughter of fire and must be led back to fire without fearing it, just as a child is led back to its mother's breast."

Some of the philosophers have called this stone of ours white copper. Lucas and Eximius say in the Turba: "Let everyone who seeks knowledge know that a tincture can only be made with our white copper. Our copper is not common, corruptible copper that contaminates everything it touches. The copper of the philosophers perfects and whitens whatever it comes into contact with." And so Plato says: "All gold is copper, but not all copper is gold. For in nature gold is similar to copper in weight and texture. But in the nature of copper there is nothing that is not in the nature of gold, as seen from its corruption in the earth, and its remaining patiently in fire and the sea. And so our copper has body, soul, and spirit, and the three are one. For spirit, body, and soul are one, because they all come out of and from one thing, and are with one thing--their root. The copper of the philosophers is their elixir made perfect and complete out of spirit, body, and soul. Therefore the philosophers have given various names to the stone, to make it evident to the wise and hidden from fools." But whatever it is called it is always one and the same, and comes from the same thing, as Merculinus says in this poem:

> The stone is hidden, and buried at the depth of a fountain.
>
> It is held cheap and cast out as if it were covered with smoke or dung.
>
> But this one living, sacred stone bears all names,
>
> as Morienus, a wise man filled with God's grace, has said.
>
> This stone is not a stone, nor something any animal could produce.
>
> Nor is it a bird, neither stone nor bird.
>
> This stone is the structure, the offshoot, and the offspring of Saturn.
>
> It is Jupiter, Mars, the Sun, and Venus; it is

Winged Mercury, and it is the Moon, alone brighter than
all others.

Now it is silver, now gold, now an element,

Now water, now wine, now blood, now chrysoline,

Now virgin's milk, now sea foam, now vinegar,

Now it distills urine in fetid bilge,

Now too the gem of salt, almizadir, general salt,

The pigment of gold constituting the first element,

Now the purged sea purified with sulphur,

Transposed in a way that they will not reveal to fools.

And so let the wise man envision this and never be
deceived,

And let what he handles never be dealt out to fools.

As Morienus says: "Our stone and the carrying
out of this teaching is like the creation of man.
For first comes coitus, then conception, then
pregnancy, then birth, and nourishment comes fifth."
Dear reader, understand these words of Morienus, and
truly you will never go astray. Open your eyes and
you will see that the seed of the philosophers is
aqua viva, while earth is an imperfect body. It is
right to call earth mother, for it is the mother of
all the elements. When the seed of mercury unites
with the earth of an imperfect body, the process is
called coitus. The earth of the body then dissolves

337

into water of the seed, and it becomes water with no division. Hali says: "The solution and coagulation of a body are two processes, but they have only one operation. Spirit cannot coagulate without the solution of the body, and when body and soul unite, each does what is like itself. For example, when water unites with earth, it tries to dissolve it with its moisture and its power. For it makes earth finer than it was before, and makes it more like itself. Water is finer than earth. The soul in the body works similarly. In the same way, water thickens along with earth and becomes almost as dense, since earth is thicker than water. There is no time difference between solution of the body and coagulation of the spirit. Nor is there any different working, and nothing that one can do without the other. That is because there is no divergent span of time in the conjunction of water and earth. It is easy to realize this if they are united or if one is separated from the other during their operations. In the same way, a man's seed is not separated from a woman's seed at the time of coition. There is one goal, one project, and one identical operation occurring to them both. So Merculinus calls coition the mixing and producing of all things."

Seeds mix together like milk and then appear mixed. Next comes conception, when the earth

dissolves into black dust and begins to retain a
little of its mercury. The masculine element is then
acting on the feminine, or azoth on the earth.
Arisleus says: "Males cannot bear young with each
other, nor can women conceive. Generation requires
both male and female. When men marry women, nature
rejoices, and true generation occurs. But if nature
is joined to an alien, unsuited nature, it cannot
produce real seed. Therefore unite your son
Gabricus, most beloved of all your children, with
his sister Beya, a lovely, gentle, and delicate
girl. Gabricus is male and Beya female. She gives
him all that she has, and even though Gabricus is
worth more than Beya, still there can be no
generation without her. As soon as Gabricus lies
with Beya, he dies, for Beya climbs on top of him
and encloses him in her womb. Nothing more is seen
of him. But she has embraced him so lovingly that
she conceives him entire in his own nature and
divides him into various parts." This is what
Merculinus says:

What was once white as milk now is changed by
blood in conception.

Pale things grow black; red and ample things
grow black.

Next comes pregnancy. The earth whitens as
water takes over, then grows and multiplies. From it

comes an abundance of new offspring. Then you must
wash the black earth and whiten it over a hot fire.
Hali says: "Take what has fallen to the bottom of
the vessel and wash it with hot fire until its
blackness disappears and its thickness lessens. Make
the added moisture blow away until it becomes an
extremely white and spotless lime. Then the earth is
pure enough to receive a soul." Merculinus says:

Pregnancy promises a space in which change can
take place.

What purgation can release is tied by bonds of
peace.

Next comes birth, when yeast unites with
whitened earth, and they become one in substance and
color. Then our stone is born to eternal life, for
then the spirit unites with the body as the soul
mediates between them. This is composition, which,
as Hali says, happens when putrefaction and
matrimonial union occur. Matrimonial union is the
mixture of fine matter with gross, thick matter. It
is also the mixture of soul and body. Putrefaction
is the process of roasting, pounding, and watering
until they are all mixed together and become one.
Then there is no diversity, no separation from the
water that has been mixed with water. And the thick
matter tries to hold on to the fine matter, while
the soul tries to contend with fire and survive it.

Also, the spirit, entire teaching." O dearest and most beloved friend, you can now easily understand obscure words because of what you have learned, and you will know that everyone agrees with them, since there is no teaching other than what I have told you. Now you hold in your power the solution of bodies and the reduction of them to first matter. Then you also have in your power their conversion into earth, along with the whitening of black earth and levitation into air. By distilling the moisture found in it, it becomes airy as it ascends, while the earth remains calcined and of a fiery nature. And you hold in your power the conversion of all these from one state to another. And you are able to increase them in a way that will be so useful to you that you will never entirely grasp it with your reason. Amen.

The Confessions of Trithemius

(The Abbot of Spanheim)

From "Traite Causes Secondes"

Translated by A. A. Wells

God is an essential and hidden fire, which dwells in all things and chiefly in Man. From this fire everything is engendered. It engenders them and will forever engender them; and what is engendered is the true Divine Light; which exists from all eternity. God is a Fire; but no Fire can burn, no Light can manifest itself in Nature without the presence of Air to maintain the combustion; thus the Holy Spirit should act within us as a Divine Air or Breath (*Ruach*), causing a breath to spring from the Divine Fire upon the interior Fire of the heart so that the Light may appear, for the Light must be fed by fire, and this Light is love, bliss and joy in the eternal Divinity. This Light is JESUS, who emanates for all eternity from JEHOVAH. Whoever does not possess this Light within him is plunged into a fire without light; but if this light is within him, then the CHRIST is in him, is incarnate within him, and he will know the Light as it exists in Nature.

All things we behold are interiorly fire and light, in which is hidden the essence of the Spirit. All things are a Trinity of fire, light and air. In other words, the Spirit (the Father) is a superessential light; the Son is the Light manifested; the Holy Spirit is a moving Breath, divine and superessential. This Fire dwells in the heart and sends out its rays all through the body and thus maintains its life. But no Light arises from the fire without the presence of the spirit of sanctity.

All things have been made by the power of the Divine word; which is the Spirit or Divine Breath emanated from the beginning from the Divine fountain. This Breath is the Spirit or Soul of the World and is called *Spiritus Mundi*. It was, at first, like air, then condensed into a nebulous substance or fog and finally transmuted itself into water. (Fr. Trans. "*The Akasha of the Hindoos*.") This water was at first spirit and life, because it was impregnated and vivified by the Spirit. Darkness filled the abyss, but by the emission of the Word, the Light was engendered, the darkness was illuminated by the Light and the Soul of the World (Fr. Trans. "*The Astral Light*") was born. This spiritual Light which we call Nature or Soul of the World is a spiritual body which may be rendered visible and tangible by **alchemical processes**; but as it is naturally invisible, it is called Spirit.

It is a living universal fluid diffused throughout Nature, and which penetrates everything. It is the most subtle of all substances; the most powerful, by reason of its inherent qualities. It penetrates every body and determines the forms in which it displays its activity. By its action, it frees the forms from all imperfection; it makes the impure pure, the imperfect perfect and the mortal immortal by its indwelling.

This essence or Spirit emanated from the beginning from the Center and incorporated itself with the substance of which the Universe is formed. It is the "*Salt of the Earth*", and without its presence the plant would not grow, nor the field become green and the more this essence is condensed, concentrated and coagulated in the forms, the more stable they become. It is the most subtle of all substances; incorruptible and immoveable from its essence. It fills the infinities of space.

The sun and planets are but coagulations of this **universal principle**; from their beating heart they distribute the abundance of their life and send it forth into the forms of the interior world and in all creatures acting about their own center and raising the forms on the way of perfection. The forms in which this living principle establishes itself become perfect and durable so that they no longer decay nor deteriorate nor change in contact with the air. Water can no longer dissolve them, nor fires destroy them nor do the terrestrial elements devour them.

This Spirit is obtained in the same way as it is communicated to the earth by the stars and this is performed by means of the Water; which serves as a vehicle to it. It is not the **Philosopher's Stone**, but this may be prepared from it by fixing the volatile.

I advise you to pay great attention to the boiling of the Water; do not let your spirit be troubled about things of less importance. Make it boil slowly, and then let it putrefy until it has attained the fitting color, for the Water of Life contains the germ of wisdom. In boiling, the Water will transform itself into Earth. This earth will change into a pure crystalline fluid; which will produce a fine red Fire and this Water and this Fire, reduced to a single Essence, produce the great *Panacea* composed of sweetness and strength - the Lamb and the Lion united.

Finis.

A Word from the Publisher

Thank you for purchasing this book from The R.A.M.S. Library of Alchemy. During his lifetime, Hans Nintzel dedicated himself to the identification, acquisition, study, retyping and, when necessary, translation of what he considered to be the most important known works on Alchemy. Hans was assisted by his sparse network of fellow Alchemists, all members of the Restorers of Alchemical Manuscripts Society (R.A.M.S.). I was an active member of R.A.M.S.

Hans provided copies of the R.A.M.S. works as photocopies. My goal is to publish all of them as professionally printed books.

The works from the original R.A.M.S. Library are republished by R.A.M.S. Publishing Company in the collection, "The R.A.M.S. Library of Alchemy," with permission of the Estate of Hans W. Nintzel.

If you have a work on Alchemy that you believe should be a part of the R.A.M.S. Library, please contact me through R.A.M.S. Publishing Company.

Philip N. Wheeler

The R.A.M.S. Library of Alchemy

The study and practice of Alchemy was extremely important to Hans W. Nintzel. He assembled this Library over a period spanning more than three decades, guided by his teacher Frater Albertus. The R.A.M.S. Library of Alchemy includes all of the most valuable Alchemical texts that Hans painstakingly located, acquired, retyped, and translated during his lifetime, with help from other R.A.M.S. members.

The following list is as yet incomplete – there are many more volumes that will be added. Some volumes have not been finalized: those are noted as TBD, or To Be Determined. Volumes that contain works from multiple authors may have only the principle author or editor listed.

Volume	Title	Author or Editor
1	Twelve Keys of Basilius Valentinus	Basilius Valentinus
2	Triumphal Chariot of Antimony	Basilius Valentinus
3	His Secret Book	Artephius
4	The Golden Work	Hermes Trismegistus
5	Three Works of Ripley	George Ripley
6	Four Works of Paracelsus	Paracelsus
7	Bacstrom's Notebooks, Part 1	Sigismund Bacstrom
8	Bacstrom's Notebooks, Part 2	Sigismund Bacstrom
9	Summa Perfectionis	Geber (Abu Musa Jabir ibn Hayyan)
10	The Five Centuries	Rudolph Glauber
11	The Greater and Lesser Edifyer	Johann Grashoff
12	Chemical Secrets and Experiments	Sir Kenelm Digby
13	The Turba Philosophorum	Arisleus
14	Das Aceton	Christian Becker
15	TBD	These volumes are reserved for the Works of Glauber.
16	TBD	
17	TBD	
18	TBD	
19	TBD	
20	TBD	
21	Alchemical Symbols, Third Edition	Hans W. Nintzel and Philip N.

		Wheeler
22	The Book of Formulas	John Hazelrigg
23	18 Short Tracts	Hans W. Nintzel
24	Bacstrom's Notebooks, Part 3	Sigismund Bacstrom
25	A Discourse on Fire and Salt	Blaise Vignere
26	The Mineral Work	Johan Hollandus
27	The Vegetable Work	Johan Hollandus
28	Lamspring's Process	Lamspring
29	The Book of Abraham the Jew	Abraham Eleazar
30	Five Short Works of Glauber	Johann Glauber
31	The Metamorphosis of the Planets	Johannes Monte-Snyder
32	Four Works of Roger Bacon	Roger Bacon
33	The Golden Chain of Homer	Homerus, Kirchweger, Nintzel, Wheeler
34	Alchemy Rediscovered and Restored	Archibald Cochren
35	Aurifontina Chymica	John Houpreght
36	The Golden Fleece	Salomon Trismosin
37	The Transmutation of Base Metals into Gold and Silver	David Beuther
38	Sanguis Naturae	Christopher Grummet
39	A Revelation of thye Secret Spirit	Giovanni Lambi
40	The Holy Guide, Part 1	John Heydon
41	The Holy Guide, Part 2	John Heydon
42	Secreta Alchymiae	Kalid Persica
43	TBD	
44	Potpourri of Alchemy, Part 1	Hans W. Nintzel
45	Potpourri of Alchemy, Part 2	Hans W. Nintzel